VD

Developments in Collection Building in University Libraries in Western Europe

Papers presented at a Symposium of Belgian, British, Dutch and German University Librarians. Amsterdam, 31st March - 2nd April, 1976

Edited by
Willem R. H. Koops and Johannes Stellingwerff

VD

Verlag Dokumentation, Publishers, München 1977

CIP-Kurztitelaufnahme der Deutschen Bibliothek

Developments in collection building in university libraries in Western Europe : papers presented at a symposium of Belgian, British, Dutch and German university librarians, Amsterdam, 31. March – 2. April, 1976 / ed. by Willem R. H. Koops and Johannes Stellingwerff. – 1. Aufl. – München : Verlag Dokumentation, 1977.
ISBN 3-7940-7020-8

NE: Koops, Willem R. H. [Hrsg.]

Druck: Hain-Druck KG, Meisenheim/Glan
Binden: Thomas-Buchbinderei GmbH, Augsburg
Printed in West Germany
ISBN 3-7940-7020-8

PREFACE

The Symposium on Developments in Collection Building in University Libraries was organized by the Dutch Group of University Libraries and held at the Free University in Amsterdam, 31st March – 2nd April 1976. In 1972 the Dutch university librarians were invited to participate in the annual meeting of the British Standing Conference of National and University Libraries (SCONUL) at Southampton. At that meeting it was my privilege to thank our hosts on behalf of my colleagues and to invite them to attend a meeting of the Dutch university libraries. Since there were also plans to invite a group of university librarians from the Federal Republic of Germany, we decided to organize a symposium of university librarians from Belgium, Germany, Great Britain and The Netherlands to discuss a theme, which should be of common interest to all of them and in which all participants should take an active part, either as speaker, as chairman or rapporteur of one of the sessions, or as participants in the discussions. For that reason too, the number of participants in this symposium was limited to fifty persons. Preparation for this meeting took up more time than foreseen, since it was not easy to find a date which did not clash with meetings of the various national library organizations.

The group of rapporteurs drafted a number of resolutions and recommendations, which were discussed and adopted at the closing meeting.

The Free University offered the conference its hospitality during the three days of this Symposium in such a way that our aims to make a number of colleagues from neighbouring countries more familiar with the structure of the Dutch university libraries, to develop personel contacts in an informal atmosphere and to discuss a problem which is recognized as one of vital interest in most countries now, were fully realized. The conditions as well as the local organization by Mr. Stellingwerff and his staff could not have been more favourable for a fruitful meeting. Furthermore this Symposium was made possible by financial contributions from the Organization for the Advancement of Information in the Netherlands (NOBIN) and from the Ministry of Education and Sciences. I would like to thank these three institutions for their substantial help in bringing our plans for a joint meeting of university librarians from four countries to fruition.

Finally we would like to thank our colleague Mr. S. M. Namenwirth, who kindly offered to read through the non-British contributions in the interest of correct English style.

W. K.

CONTENTS

INTRODUCTION

University Libraries in The Netherlands

W.R.H. Koops

Before introducing to you the theme of this symposium, I take this opportunity to give you a brief sketch of the structure and organization of Dutch academic libraries while explaining to you some of their characteristics and traditions.

In The Netherlands the university library is completely dependent upon the university to which it belongs; furthermore, according to the Higher Education Act of 1960, the universities are self-governing. They have the competence to regulate their own affairs, within certain limits laid down in this Act, and this in accordance with four-year plans to be approved by the Ministry of Education and Sciences. Moreover, the university budget has to be presented every year to the Minister for approval.

As a consequence, cooperation between university libraries take place on a completely voluntary basis.

From a legal point of view the thirteen Dutch institutions of university standing can be divided into two main types:
1. State universities and
2. private universities.

1. State universities are established by the government. They include the old universities: Leiden (1575), Groningen (1614), Utrecht (1636), the three technological universities: Delft (1905), Eindhoven (1956), Twenthe (1964), and finally the Erasmus University of Rotterdam (1973) and the University of Limburg at Maastricht (1975). To these institutions can be added the Agricultural University at Wageningen (1918), which falls, however, mainly under the competence of the Ministry of Agriculture.

2. Private universities are founded by societies, generally on a religious basis. They include the Calvinist Free University at Amsterdam (1880), the Roman

Catholic University at Nÿmegen (1923), the School of Economics at Rotterdam (1913, which amalgamated with the School of Medicine and became the Erasmus University in 1973) and the Roman Catholic School of Economics at Tilburg (1927).

The Municipal University of Amsterdam (1632) holds an intermediate position insofar as, although being a public university, its status is more like that of a private university.

Personnel and operating costs, as well as construction investments of the state universities, are fully met by the central government. Originally the private universities were self-supporting but now they also receive a 100% reimbursement. As to the financing modalities for the private universities, according to the Higher Education Act (art 96 ter), "their special character" has to be taken into account, a phrase which gives them somewhat more liberty of action. We all know that some are more equal than others.

The number of students has been growing rapidly over the last twenty five years and presumably will continue to grow, although at a much lower rate, because of the present economic recession which will apparently continue in the years to come.

The solution to the problem of overcrowded universities has not been sought in establishing new universities, but rather in expanding the existing ones by creating new faculties and by enlarging facilities.

On December 1, 1975 the total number of students was 122.785. This number is spread over the individual universities as follows:

Leiden	13049	Wageningen	4291
Groningen	13794	Nijmegen	13435
Utrecht	19296	Tilburg	3577
Amsterdam (G.U.)	19814	Eindhoven	4317
Amsterdam (V.U.)	11446	Twenthe	2158
Delft	10250	Limburg	99
Rotterdam	7349		

In The Netherlands the director of the university library is at the same time the university librarian. As such he is responsible for all books and periodicals within the university. The organization of library provisions is not identical in all Dutch universities. The older universities show the traditional system of a central library together with a number of departmental libraries, mostly well over 100, of varying size and quality. Nevertheless, the library is considered to be an entity. As a consequence, with a single exception, the ordering and cataloguing of all books and periodicals is carried out in the central library and there is a central author/title as well as a subject catalogue of the university's total holdings of books, periodicals and related materials. The budgets of the various departments include the expenditure for their libraries. These budgets are allocated by the university through the faculties. More often than not the university librarian will not be consulted. The total expenditure of departmental libraries is often twice, or even more that, of the central library. The departments are free to decide how to spend this money and coordination can only be practised on a voluntary basis.

In the new universities the situation is arranged more conveniently. There the university librarian is also in charge, or at least in control, of the departmental libraries, which are mostly on the faculty level.

The national library – the Royal Library at The Hague, established in 1798 – exercises a central function in the Dutch library system in that it houses the Union Catalogue of Books *(Centrale Catalogus,* C.C.) having 94 participating libraries and the Union Catalogue of Periodicals (C.C.P.) having 264 participating libraries. In the latter catalogue are also included the holdings of government and industrial research libraries. It is through these catalogues that our interlibrary lending system operates, a system which has always been very liberal, thanks primarily to the voluntary cooperation of the university libraries.

It should be mentioned here that in addition to these union catalogues the library of the Technological University at Delft keeps a Central Technical Catalogue (CTC), listing in particular the holdings of industrial libraries.

As I have said before, consultation and cooperation between Dutch libraries is on a voluntary basis. In 1922 the State Advisory Committee on Library Affairs *(Rijkscommissie van Advies inzake het Bibliotheekwezen)* was established to advise the government on all library matters, although public library matters were in practice excluded. Under the chairmanship of the chief librarian of the Royal Library, it included the chief librarians of all academic libraries. About ten years ago it extended its field of interest to all library affairs, and a number of representatives from public libraries were appointed to its membership. This committee functioned not only as an advisory body but for mutual consultation as well. Over the years its importance grew considerably and it also assumed executive tasks. It appointed several subcommittees to study special problems, e.g. new cataloguing rules, mechanization of acquisitions, manuscripts and rare books, audio-visual materials, library building.

From the numerous reports submitted by the committee mention should be made here of a long-term programme for the development of research libraries in The Netherlands (1969) and a plan for the coordinated acquisition of foreign research literature (1974), to which I will come back later on.

Together with a Bill on Public Libraries, a Bill on the Library Council was adopted in 1975. Its effect was to replace the State Advisory Committee on Library Affairs by a new Library Council. The new Council is explicitly an advisory body, and in this respect it has a more limited field of activity than its predecessor. For its executive tasks a new umbrella has to be found. This could be the Dutch Association of Librarians, but also an association of university and research libraries. Under the State Advisory Committee, the university libraries were united in a subcommittee. At present there is a loose organization of the libraries of the original six, now eight universities, the Convent of University Librarians in The Netherlands. At the moment the question is under discussion whether this association should, in order to strengthen its organizational structure, be extended to include all university libraries, as well as the Royal Library. A decision will be taken in the near future.

We hope to be able to create an effective body, which not only will provide opportunities for the discussion of matters concerning this category of libraries, but can also represent their views to outside agencies, undertake common projects for all or most of its members and apply for the money necessary to carry out its programmes. This is not the place to draw up a programme of activities, but I am convinced that cooperation in collection building in university libraries would be on this list.

In most countries library cooperation is an issue of current interest, on which the attention not only of university librarians and users, but also of the administration is focused. On the one hand we have seen a tremendous growth of printed and non-printed materials as well as a rapid expansion of research needs. On the other hand it is evident that we are entering a period in which funds will be diminishing and difficulties will be growing. The first step will have to be cooperation within one university and this should be made a major objective in every institution. In most Dutch universities there is a tendency now towards more coordination. Nowadays it is acknowledged that individual libraries can no longer meet the demands made upon them by their users and that, as a consequence, a sharing of resources will be necessary. Several ways are being considered to improve the access of the users to the relevant information, but often conflicting interests interfere with the most appropriate solution. It is, however, clear that through automation this process will be speeded up; by storing bibliographical data in a computer and by providing direct access to them, thus facilitating central cataloguing and improving the interlibrary lending system.

The prime question is how to get needed materials to the libraries. In The Netherlands, the State Advisory Committee on Library Affairs presented in 1974 the draft of a plan on the coordination of the acquisition of supplementary foreign research literature. The underlying idea of this plan is to ensure that the major foreign research literature in all relevant fields will be present in a Dutch research library and will be available to the users. The committee has opted for cooperation rather than for centralization. The plan has been submitted to the Minister of Education and Sciences and is at the moment under critical discussion in the Academic Council, within the universities and of course among librarians. Different countries have taken different approaches to this problem, but it is a common problem to all of us and we hope to benefit from your experience.

This is the reason why we considered it an appropriate theme for this symposium. We do hope that the discussions of the next few days will be fruitful, will contribute to further improvement in library service and will hence benefit our users.

A. COOPERATION IN COLLECTION BUILDING WITHIN A UNIVERSITY

Cooperation in Collection Building in British Universities

Kenneth Garside

It is my privilege to read the first paper at this joint meeting in Amsterdam in which we British librarians representing SCONUL are delighted to have the opportunity of joining with our Dutch hosts and our colleagues from Belgium and Germany to discuss developments in collection building in university libraries. My allotted theme is that of cooperation in collection building within a university, and in being the first to speak I am at a disadvantage in that I do not know how subsequent speakers propose to deal with the subject. I am very conscious of the differences there are between the pattern of university library provision in Britain and that in, for instance, Germany; unfortunately I am not sufficiently familiar with the Dutch and Belgian systems to draw precise parallels with them, though Mr Koops's introductory remarks suggest that the Dutch position represents a compromise between the British and German systems.

I think I should therefore begin by stating that the normal pattern of library provision in a British university, whether old or new, is that of a single unitary library under centralised control. For geographical reasons, when for instance there is a large campus or where the university occupies more than one site, the library may be divided between different buildings, usually on a faculty basis, but it remains organisationally a single university library under a single librarian or director, and governed by a single library committee concerned with matters of policy. Independent departmental libraries, like the seminar and institute libraries of German universities, are almost unknown in Britain. Only in Oxford and Cambridge, and in London, is the pattern different for historical reasons; I shall refer to these again later, but for the moment it will be more useful for me to confine my attention to the unitary libraries which are characteristic of the typical British university.

Since there is normally a single library organisation in a British university, it follows that there is rarely any question of cooperation *between libraries* within the university; but there is considerable scope for cooperation between the teachers of the university and its library. The British university is a close-knit community in which it is possible for the library to establish and maintain close touch with the various academic departments in its primary rol of building up its collection to meet the specific needs of the teaching and research carried on in every subject represented in the curriculum. This liaison with the academic departments is facilitated formally in many university libraries by the organisation of the senior staff as subject specialists, rather on the lines of the German *Fachreferenten* but, instead of working largely on their own in matters of book selection, they enjoy a positive and continuing relationship with the teachers of the subjects for which they are responsible. The teachers constantly recommend new titles for addition to the library in their own subjects, not only for the use of their students but also in their special fields of research, thus ensuring that the immediate needs of the academic departments are taken fully into account in building up the library's book stock. Moreover, the librarian and his senior staff have opportunities for regular *in*formal contact with professors and lecturers in the senior common room, or at lunch in the refectory, where library matters can be discussed in a relaxed social atmosphere. The teaching staff as a whole is represented on the library committee which advises on university library policy.

This close relationship between the library and the academic departments has had its effect on the physical disposition of the library's collections. These are often arranged on a subject basis, corresponding to the academic departments themselves, with separate reading rooms set apart for each subject (e.g. French, or chemistry) or group of related subjects (e.g. Romance studies, or the physical sciences). Since there is only one library serving the university it has become a general principle to include in such subject collections both the textbooks and monographs prescribed for reading by undergraduate students and at least the more general research material in the subject. As many books as possible are made available on open access and arranged on the shelves in classified order, so that the student is able to browse widely among the literature of his subject and broaden his intellectual horizons beyond the texts which he *must* read for his course. He is also encouraged to use the library's collections in subjects other than his own.

But lack of space often prevents the British university library from achieving this ideal solution of placing all the books it would wish on open shelves. Specialised research material will therefore often have to be kept in closed bookstacks, though usually with facilities for teachers and research students to have direct access to it. And some university libraries have broken away from the tradition of keeping undergraduate and research material together by setting up separate reading rooms, or indeed separate library buildings, for the use of undergraduate students, the shelves there being provided with the books such students need for their courses. This development may have been dictated by problems of space in the main library, particulary following the enormous increase in student numbers since the war; but it is a development which I personally consider retrograde since students are thereby passively discouraged from browsing in the main collections in their subjects – though they are not, of course, prevented from doing so because they continue to have access to the collections of the main library whenever they wish to use them.

I make no apology for digressing into questions of the organisation of our university libraries in this way, because it has helped to show that our libraries embrace within themselves the functions of different types of library in other countries. In particular, the organisation of our university libraries by subjects, with large numbers of books on the open shelves, almost all of them available on loan for home reading by staff and students alike, combines in a single unit the functions of the central library and the seminar and institute libraries of the traditional German university system.
I would, however, draw attention to one area in which there is a form of cooperation which leads to the development of university libraries as centres of excellence in particular subject fields. This is cooperation between local learned societies and a university library, whereby the university library's resources are enriched by the deposit of the private libraries of learned societies. Such collections usually come to universities when the societies themselves cannot afford either the space to house them or the funds to administer them properly, and their deposit in the university library brings significant benefits to both parties: the contents of the society's library enhance the resources of the university library in the particular subject field, especially in its range of periodical literature, not only by introducing new titles but by keeping runs of periodicals up to date by the accession of the journals received in exchange for the society's own published proceedings; the university library's readers have access to the material in the society's collection; and the society's members not only have continuing access to their own books and periodicals but also to the entire book stock of the university library as well. Perhaps the most obvious example I can quote in this context is the library of University College London, which has acquired in this way the libraries of a dozen learned societies, in the humanities as well as the sciences, and which consequently has one of the most comprehensive collections of scholarly periodicals in London, if not in Britain, after the British Library.

I have already referred to the fact the Universities of Oxford, Cambridge and

London are exceptions to the general rule of library provision which I have sought to describe. Oxford and Cambridge – the two oldest English universities – each embrace a number of colleges which derive from a monastic tradition and are primarily residential communities rather than teaching units. Each college has its library which makes provision for undergraduate needs and which also usually has a considerable collection of older books collected over the centuries. Each of these universities also has its central library – the Bodleian at Oxford and the University Library at Cambridge – which enjoys the right of legal deposit under which it receives free of cost currently published British books and periodicals and has built up over the years a vast collection of research material. These central libraries are therefore principally research libraries, though the Bodleian at least has one reading room devoted to the needs of undergraduates. Library provision in support of teaching is, however, based mainly on faculty libraries which, as their name suggests, are devoted to the literature of particular specialised fields. Oxford and Cambridge can therefore be said to have a three-tier system of library provision, with the college libraries at the grass roots, the faculty libraries in the middle, and the central libraries as the main research collections at the top. Following the recommendation of the Committee on Libraries of the University Grants Committee in 1967(1) that library provision in complex universities should be investigated with a view to achieving some degree of rationalisation as between the many libraries in a single university, both Oxford and Cambridge set up committees to examine the problem in relation to their own universities. Oxford has certainly produced a substantial report on the subject(2), and I am only sorry that our SCONUL delegation here does not include any colleague from either university who can describe the outcome of these studies.

The University of London is different again from Oxford and Cambridge in that it is a *federal* university – not so much a single university as an amalgam of a number of universities. It is therefore very much *sui generis,* and its library problems are unique in Britain. The special character of the University of London derives from the fact that it started with the foundation of two rival colleges – University College and King's College – in the 1820s, each of them with the characteristics of, and indeed intending to become, separate universities. The University of London, as such, was founded ten years later as a purely examining body because the government of the day could not tolerate the idea of having more than one degree-awarding university in a single city. Times have changed since then! During the intervening years more colleges and medical schools have joined the university, some being actually older foundations than the University itself, and there are now fifteen colleges, mostly general and embracing a wide range of faculties, but others, like the London School of Economics and the School of Oriental and African Studies and others confined to the faculties of sciences or medicine, more limited in their subject coverage. Each of these colleges and medical schools is virtually self-contained, and the larger ones certainly have the academic and administrative machinery to enable them to function as fully-fledged universities if the University of London as such were swept away overnight. Yet all these institutions teach for common degrees and are funded from a common source. They are therefore dependent upon an enhanced central administrative organisation which not only examines and awards degrees, as was its original function, but now also provides the means of consultation between the teachers in the different colleges for determining the content of courses, as well as dispensing the grant made by the University Grants Committee for the University as a whole. More recently there have been established a number of postgraduate institutes each devoted to a particular subject field, in which the teachers from the various colleges collaborate in research activity; and a number of specialised medical postgraduate institutes has similarly been created. Each of these colleges, medical schools and postgraduate institutes has its own library; there are indeed more than sixty libraries in the University, with a total book stock of some five million volumes. For the historical reasons at which I have hinted, all these institutions remain virtually autonomous, zealously guarding their independence and very suspicious of the intentions of the central machinery of the University.

I have digressed into the historical development and organisation of the University of London to illustrate its complex nature, which is inevitably reflected in its pattern of library provision. There has emerged in effect a two-tier library system. At the grass roots we have the libraries of the colleges and medical schools, which provide for the teaching and much of the research carried on within their own institutions. Then we have the central University Library – now the largest, with a million volumes – and the libraries of the central postgraduate institutes. All teachers and students belong to one of the colleges or medical schools, and therefore have the right to use their libraries. All members of the University also have the right to use the University Library, while academic staff and research students have access to the institute libraries in their subjects, where such institutes exist; they do not, however, cover all subjects, and in particular there is no central research provision for the pure sciences and technology.

I have emphasised the independence of the colleges in London. The result of this parochial attitude is that there has until recently been no formal cooperation between the libraries of the various colleges, nor between their libraries and the University Library and the libraries of the postgraduate institutes, save in one or two specialised fields such as law — and even in this subject cooperation has largely lapsed because the number of law students in each college teaching the subject became so large that each institution felt the need to cover the whole subject in its own library. There are similarly no union catalogues of the holdings of the different libraries in the University to enable members of one college to know which titles are held in any other college library, and in any case they would probably not have right of access to them, other than as inter-library loans, if they did know of their existence. One can therefore describe the library situation in the University of London as one verging on anarchy.

However, following the publication of the report of the Committee on Libraries of the University Grants Committee, to which I have already referred in the context of Oxford and Cambridge, the University of London set up a Committee on Library Resources to examine its library problems and to make appropriate recommendations. This Committee included in its membership heads of colleges and institutes, teachers in a wide range of subjects, and librarians. Its report, running to 226 pages, was published in 1971.(3) While recognising the special interests and aspirations of the colleges, it recommended far-reaching measures towards the rationalisation and coordination of library resources throughout the University, its principal recommendation being the establishment of a permanent central body charged with the task of organising such measures. The Committee's report was referred to colleges and other bodies in the University for comment, and it received a degree of unanimity of approval which was remarkable in an academic community. As a result the Senate of the University established in 1973 a central body called the Library Resources Coordinating Committee with, broadly, the task of implementing the recommendations of the Committee on Library Resources. The membership of the new permanent body includes not only the Vice-Chancellor and the Principal of the University but also heads of colleges and institutes, teachers, and librarians. Many other teachers and librarians have been drawn into the detailed work of its various sub-committees and advisory panels. The Committee has set up a coordinating staff under a Director of Central Library Services who is also Director of the University Library — the post which I now occupy.

The setting up of the Library Resources Coordinating Committee coincided with the worsening of the financial situation — a circumstance which has given it a good start since college and institute libraries began to realise the benefits which could accrue to them from the existence of centralised coordinating machinery. The financial situation was indeed to dictate the priorities of action. The shortage of funds available to individual libraries, coupled with the rapid escalation of book prices and more particularly periodicals subscriptions, led to the first priority being given to the compilation of a union list of periodicals taken in the various libraries of the University. This list is now being compiled with the aid of a computer, and when it is completed it will enable individual libraries to decide which periodicals they can cancel in the knowledge that they are available somewhere else in the University. It is intended that the union list of periodicals will be followed by the production of union lists of current accessions to the libraries of the University, also computer-based and so organised that any library considering the purchase of a particular title can interrogate the data base to ascertain whether or not some other library is already acquiring it; this will serve to obviate unnecessary duplication in the purchase of monographs. Ideally this project will be followed by the compilation of a computer-based union catalogue of the books already in the libraries of the University and so facilitate the discarding of unnecessary little used material, but we have to recognise that this will be an expensive project since there are some five million volumes, the vast majority of which are too old to have MARC records. All these measures will contribute to the rationalisation of collection building throughout the University.

In the meantime we are setting up sub-committees to study problems of library cooperation in particular subject fields. These sub-committees consist of teachers nominated by boards of studies (the central University bodies which bring together the teachers in each subject from the various colleges to determine academic policy relating to their subject), together with representatives — usually the subject specialists on the library staff — of those libraries which have significant holdings in the subject, and, perhaps of particular importance since it carries the principle of cooperation in collection building outside the University, representatives of other learned libraries in London concerned with the subject; in most cases the Reference Division of the British Library is invited to collaborate in the work of these subject sub-committees. It is hoped that their deliberations will lead to a real degree of rationalisation of library provision in London.

Perhaps at this stage I ought to say a word about the responsibilities of the various libraries in the University in collection building. It is generally accepted that college and medical school libraries shall provide on their own shelves the literature needed by students in connexion with the courses they are studying. Research provision is, however, more complicated. In subjects where there are postgraduate institutes it is the responsibility of these institutes to collect research material in their own fields. Some subjects are taught in one college only, and in such cases that college library undertakes the responsibility for collecting research material; leaving aside such obvious examples as the London School of Economics and the School of Oriental and African Studies, I can quote the example of King's College which, alone in the University, teaches modern Greek literature and history and Portuguese studies, and therefore possesses the major collections in these two subject fields. Those subjects for which there is no postgraduate institute and which are taught in more than one college are covered by the central University Library, which also provides additional undergraduate material in all subjects. This undergraduate material in the University Library is intended to serve as a second string for those students who find the books they want already borrowed from their college libraries, and also to offer a wider range of books in each subject for the benefit of students from the smaller colleges whose libraries are less well equipped in this respect. To this end the University Library is steadily developing, as space permits, a series of subject reading rooms in which both undergraduate and research material are shelved together to form browsing collections, while it has also adopted the policy of confining to the Library as reference material books in demand by students and periodicals, so that the visitor from an outlying college can be reasonably certain to find the material he needs on the shelves and not on loan.

To help libraries carry out a rational programme of development of their collections within the framework I have outlined the Library Resources Coordinating Committee has established a central book fund from which grants, normally amounting to half the cost, are made to individual libraries to enable them to purchase particularly important and expensive works in the subject fields for which they have a special responsibility, or towards the cost of developing resources in a particular subject field within a policy agreed by the appropriate subject sub-committee.

But any rationalisation of library resources is, of course, useless unless the members of one college can have access to the libraries of the others, which is at present not usually the case. We are therefore seeking to liberalise access to libraries, and as a first step we have devised a reader's ticket which will allow a member of one college to gain access occasionally to the library of another to consult particular material of importance to him. It is also a condition of the grants from the central book fund that any member of the University needing to have access to material so purchased shall be allowed to consult it, whether or not he has any right otherwise to use the library which has acquired it.

The activities of the Library Resources Coordinating Committee go beyond those I have outlined – into such areas as the establishment of a central information service, the management of manuscript and archive collections, and matters concerning library staffs and their further training – but as these do not fall within the theme of this symposium it would be irrelevant for me to go into details here.

There is, however, one cooperative activity which antedates both the Committee on Library Resources and the Library Resources Coordinating Committee, but which has a bearing on cooperation in collection building if only in a negative sense. As long ago as 1961 a Depository Library was opened to provide accommodation for little-used books and periodicals for which individual libraries in the University no longer had space in their own premises on costly and congested sites in central London. This Depository Library, situated in the grounds of one of the outlying colleges some thirty-five kilometres from central London and therefore occupying a relatively inexpensive site, serves two purposes: it provides space in which libraries can deposit temporarily little-used material which they cannot accommodate until new library buildings or extensions become available; and, more importantly, it collects older books and periodicals no longer required in individual libraries but which it is considered should remain available for consultation somewhere within the University. Although the existence of this latter common store implies a degree of impoverishment of the resources of individual college libraries, it represents a major aspect of cooperation in that there is now developing in the Depository Library a significant collection of older material which might otherwise have been discarded and lost to the University. Books and periodicals in the Depository Library are available on loan to any library in the University, and also as inter-library loans to any other library in Britain, or indeed abroad, through the Lending Division of the British Library.

The Depository Library, which is now part of the Central Library Services under my control, was indeed the first concrete example of library cooperation within the University of London. For twenty years, however, there has been a Standing Conference of Librarians of the Libraries of the University of London, based on the model of SCONUL, which has served as a forum for the discussion of library problems in the University. Since the establishment of the Library Resources Coordinating Committee this body has been embraced within its organisation as the Advisory Board of Librarians, and all kinds of problem relating to library provision and cooperation have been referred to it by the parent committee. It is clear that this Board will have a considerable part to play in all matters relating to cooperation in collection building in the libraries of the University.

If I have concentrated particularly on the situation in the University of London this is not only because it is the area with which I am naturally most familiar, but also the area which offers most scope for cooperation within a single university in Britain and therefore the one whose problems are most relevant to the theme of this symposium. But in my portrayal of the picture both of library provision in Britain generally, and that of London in particular, I have necessarily had to over-simplify, for every university in Britain is an independent institution and even where institutions appear superficially to be similar there are many underlying differences. In a paper of this length it is impossible to be comprehensive.

References

1. University Grants Committee: **Report of the Committee on Libraries,** London, HM Stationery Office 1967.
2. University of Oxford: **Report of the Committee on University Libraries** 1966.
3. University of London: **Report of the Committee on Library Resources** 1971 (SBN 7187 0213 1).

Coordination of Collection Building within the Dutch Universities

J. R. de Groot

The theme of this conference is the coordination of collection building, both within universities and on the national level. I have been asked to speak to you about the coordination within Dutch universities.

For practical reasons I will limit myself here to the situation in which collection building activities are shared by the central library and a number of institute libraries.

The libraries in the older universities have a long history. Up to about the middle of the nineteenth century meeting the demands of university research and teaching did not create too many problems. The number of students was small. The teaching consisted mainly of formal lectures; the professor would speak from a lectern to the students who were supposed to be listening, not discussing. The curriculum was fairly constant and so was the demand for books.

The instructional programme was concentrated in a few buildings. All book facilities were housed in a building near to the lecture rooms. Thus one central library fulfilled all needs. However, after ca. 1870 the teaching began to change, mainly because of developments in the study of the natural sciences. Lectures were more and more replaced by seminars, tutorials and practical work, resulting in a different use of books and a need for more library facilities.

The number of students has grown enormously since then. The old towns which had harboured the universites right in their city centres, could hardly accomodate this spreading growth, so university facilities became more and more dispersed over the cities. One central library was no longer sufficient to meet the needs of modern teaching and research for so many students. Thus book collections grew up within the various departments and laboratories.

In this manner the typical situation arose of a central library in the old centre of the town together with numerous institute libraries spread out over a large area, having various degrees of independence.

This is the situation as we know it to-day in most of the older universities. At some of the newer universities the situation is slightly different. Some have been able to avoid this library problem, for others it has not become a real problem yet, because of their limited size.

To give you a more detailed pictue of the Dutch situation, I shall take a closer look at some aspects of the relationship between the central library and the institute libraries. To this purpose I refer to two questionnaires on the subject of institute libraries, which were used for a survey by the State Committee for Library Affairs. The first questionnaire was composed and distributed in 1969. Unfortunately the data collected were not studied till 1973. At that time a few more up-to-date data were added. The final report has been finished recently, but is not yet available for publication.

In 1973 a second questionnaire was put together in order to get a better idea of the actual situation in the small units. It was sent to a sample of 45 institute libraries in 6 universities. As this is only a 5 % sample, the results may not be very conclusive. However, the report does give us some added information.

For the purpose of this discussion we will define "institute library" as it was defined in the questionnaires: "An institute library is any concentration of books and periodicals outside the central library, which is not controlled in every respect (i.e. personnel, material and budget) by the university librarian."

To begin with we can divide the universities in three groups. A group of six universities having a central library and between 90 to 150 institute libraries. A second group owning between 50 to 70 collections outside the central library. And a third group of three universities possessing less than 30 institute libraries.

As far as organization is concerned, as a rule there is no hierarchical relationship between the university librarian and library personnel of the institute libraries. Sometimes the advice of the librarian is asked in relation to the appointment of new personnel, but only two out of the twelf librarians can make any demands.

As for finances, there is some variation in the extent of independence which the institutes have within the various universities. In most older universities at least a large number of institute libraries control their own book allocations. The institute usually decides which part of its budget is to be used for library purposes, although the actual degree of independence of the various institute libraries varies. Sometimes the central library knows the amount of the institute's allocation, but this is certainly not always the case.

The smaller libraries tend to have more control over the budget of the institutes than the larger ones. Only in a few universities is the budget officially controlled by the central library, at least in theory. The control of the budget ties in directly with the control of selection and acquisition. If the budget is controlled, the selection and collection building tend to be under some control as well.

In most of the large libraries there is no official control on selection and no formal division of subject areas among the various library units. Fortunately there are often some voluntary arrangements whereby the institutes are expected to limit themselves to their own subject areas, and whereby the central library refrains from buying to any great extent in certain specialized fields.

Although the central library in most instances checks all orders to see if the book is not already available elsewhere in the university, its power to do anything about it is usually very limited. It can warn the originator of the request and demand a reason for the duplication, but in most cases the final decision lies with the institute, and not with the university librarian.

The central library is usually responsible for the actual acquisition of most of the books. For books thus ordered and received by the central library, normally catalogue descriptions are made centrally.

However, institute libraries often have direct contacts with booksellers also. Such contacts are mainly used for obtaining books on approval, either at the request of the library, or at the initiative of the bookshop. In the opinion of the central libraries, institute libraries keep a large proportion of these "on approval" supplies, varying from 25 to 100 %.

The answers from the sample of institute libraries show that on average 50 % of these supplies is actually bought, comprising about 33 % of the total acquisition of the institute libraries. These direct purchases are of course not checked for duplication with other library units at all, and are not always sent to the central library for cataloguing. Thus they remain completely outside the collection control of the central library and may therefore not be recorded in the central union catalogue. Exchanges constitute another source of book supplies. Most of the institute libraries maintain exchange relationships with other organizations, up to an average of 85 such contacts. The material thus acquired is usually a mixture of brochures, reports, dissertations, periodicals, etc., which are indeed not always suitable for recording in the central union catalogue. It is therefore often catalogued in the institute library. The central library usually has no control over this kind of acquisition, and often no information about it either. Virtually all institute libraries do their own selection of material. University staff do sometimes make suggestions, but the main part of the selection is done by library personnel.

There is usually a part-time subject specialist responsible for this job, but about a third of the sample institute libraries stated that the non-graduate assistant-librarian also helped with the task of selecting materials. So far the findings of the report.

It is obvious that the task of coordinating the total university library collection development programme is made very difficult, if not impossible, for the university librarian. Fortunately some of the newer libraries have learned from the example of the older ones, and tend to have a better organization. In the upcoming discussion it might be interesting to ask our colleagues for more details about their own situations.

Against the background of this general survey, I would like to tell you something about my own experience in the area of coordination of collection building in Leiden. As you may know, the Leiden University Library has a very long tradition of book-collecting (namely four hundred years) and a great. amount of decentralization (115 institute libraries). However, there are also the beginnings of integration in the form of completely centralized ordering, cataloguing and classification.

Categorized by the amount of coordination among central library and institute libraries, one can distinguish three groups of faculties in Leyden:

The first group includes the faculties of theology, languages and law. Between the central library and these faculties exists a very satisfactory kind of coordination, based on a long tradition of book collecting. The collections in these subject field are mainly centralized and there are very few institute libraries. Of the institute libraries that do exist, those for the Faculty of Law are in fact branches of the central library. The faculties of theology and law are consulted regularly on their library affairs. A similar arrangement is being prepared for the language faculty, also in view of the expected new library building.

The second group is formed by the faculties of philosophy and social sciences. The coordination with these faculties is less satisfactory. However, in both cases we can perceive active attempts towards an improvement of the situation. In the case of the faculty of philosophy this has recently resulted in a new division of tasks, whereby the responsibility for the collection building in that subject field has been handed over to the university librarian.

The social sciences have developed strongly after the second world war and so has the need for books in this field. This growing demand was met in a rather uncoordinated fashion by a number of institute libraries. The central library had no tradition of book-selection in this area. In the middle of the nineteen-sixties, a start was made to improve the situation by appointing a subject specialist for this field in the central library. Since then, this specialist has built an integrated study collection for the social sciences in a reading room in the central library and has had the responsibility for the development of a balanced central collection, as a counterpart to the decentralized collections. Avenues for consultation between the central library and the faculty, and between the various institutes have yet to be developed.

The third group consists of the faculties of medicine and science. The coordination between the central library and these faculties has hardly developed at all.

The relationship with the faculty of medicine has the best chances of developing favourably. Since the nineteen-fifties, there is a central medical library, started at the initiative of my predecessor. On account of various incidental problems, this library has up to now functioned mainly in the field of lending, and has not yet played an effective part as an agent for the coordination of collection building.

Of the means and techniques employed by the Leiden University Library for the coordination of collection building, I would like to mention three:

First of all the budget. In the last few years we have made considerable progress in our attempts to create one library budget for the whole university. In these matters an important role has been played by the Committee of Advice for the Library Budget, which consists of the deans of faculties, with the university librarian as chairman. However, the book allocation of the faculties of medicine and science are not yet integrated in this library budget. In the beginning of this year the university asked me to make a detailed study of the possibilities for a total integration of all book allocations into a single library budget.

The present library budget has been divided with considerable accuracy into ca. 70 subject areas. We keep a check on the actual spending to make sure that the allocation for each subject is not exceeded. The experience gained from this detailed division of the budget is very favourable, as long as the distribution is not applied too strictly.

It certainly is conducive to the building of a coordinated well-balanced collection.

The second means is the central ordering department. In principle all institutes are expected to send their orders to the central library for central processing. These orders are always checked, to see if the book is not already available in, or on order for, another library in the university, and to make sure that the institutes do not buy materials which are too far removed from their subject areas. If either of these is the case, the originator of the order is notified and asked to justify the duplication. The central library can only suggest withdrawal. In reality, ca. 25 % of these potential duplicates are cancelled in this manner.

I consider this a good result considering that we are dealing with a voluntary arrangement.

The third technique is the use of a collection profile. Within the central library we are working on a project trying to define exactly who buys what in which field and why. We are trying to fit these data into diagrams, with textual explanations. Hopefully this will give us a clearer picture of what is actually happening, how new acquisitions fit in with the present collection and which areas are neglected. We would like to apply the same technique to the total university book collection. This would give us a good overall picture and might enable us to arrive at a balanced total library collection.

Summarizing we can say that the coordination of collection building within Dutch universities is still in an early stage of development. Yet, if we want to maintain, and improve, the quality of our collection building efforts, it will be necessary to stimulate energetically the present movement towards integration. In my opinion, universities will have to aim for the realization of the following three preconditions: Firstly, there must be one integrated library system per university, in which collection building is organized according to binding arrangements or regulations, determining which units are allowed to buy what kind of material in what subject fields.

Secondly, one integrated library budget is needed. This must be divided in such a way that it supports and underlines the above-mentioned agreements on subject coverage.

Thirdly, there should be one person or body having the authority to supervise the strict application of the first two conditions and having the power to take decisions in this area.

Acquisition Policies and Methods in a New Graduate College in Belgium

H. D. L. Vervliet

The *Universitaire Instelling Antwerpen* (UIA) was created in 1972 to provide for graduate studies at Antwerp; it was preceded for about a decade by undergraduate studies at the other two university colleges at Antwerp, namely the *Rijksuniversitair Centrum* (about 1600 students) and the *Universitaire Faculteiten Sint-Ignatius* (about 2600 students). At present the UIA has about 1500 students and 300 teaching staff. These numbers will probably rise to 5000 and 600 respectively, by 1985.

A fusion of the different scientific library collections at Antwerp in the sense of the well-tried German *Stadt- und Universitätsbibliothek,* however desirable, was not feasible for both political and topographical reasons. The non-integrated campuses have necessitated a separate UIA-library. To UIA, this means a serious extra but inevitable effort. At present, the acquisition of books and periodicals absorbs about 3.5 % of the overall university budget. Total library expenses, including wages, amount to 7.8 % of the total university budget of £ 6,000,000 (in 1975).

UIA is being built on a large campus (200 ha) on the outskirts of town. Its academic structure is based on the Anglo-American concept of departments. As such it is quite unique in Belgium. It does not have faculties, no chairs, no institutes. Much stress is put on collegial decision-making.

Obviously the library organization has to fit into this general structure. Conspicuous is that there are no institute libraries and that there is no central library. The library consists of a network of departmental libraries. Their location, their book-collections in open access (exceptional in continental Europe), their long opening hours all intend to link the library firmy to departmental teaching and research activities and to promote selfactivity on the part of the students.

The negative aspects of the scattering of the book-collections over half a dozen locations on the campus are real. They have been enumerated many times.(1) Therefore the decentralized concept is counterbalanced by the fact that all departmental libraries are organized in a uniform and consistent way by a central logistic unit. This unit is not called "central library" but

"Library Centre." The balance between centralization and decentralization works as follows: Funding (yearly) and selection (continuous) are primarily departmental responsibilities, the Library Centre maintaining only an overall control. Under specific control of the Library Centre are acquisitions, fund control, cataloguing and reader services in the different reading rooms.

A small amount of the budget, viz. maximum one tenth, may be "privatized" for books and periodicals to be used in laboratories, clinics, or for a researcher's private needs. The fact that during the last three years this "private" book-budget has only been used to an extent of 6 % (instead of the allowed 10 %) shows that the reading and documentary needs of individual scholars have been met by the departmental libraries.

Funding originates with the departments. Every year a book-budget is proposed by each department. Books and periodicals must therefore compete with other priorities. Each department, through a collective decision, fixes its priorities and makes known its needs about instruments, administrative expenses, travelling, etc. The library fund is subjected to a kind of competitive market place mechanism. It may sound surprising but up to now all departments, including the scientific ones, have given the funding of their departmental libraries a very high, sometimes the highest priority. Even though some book-funding proposals had to be graded down (being too high), the book-funding budget (net for acquisitions, viz. excluding all staff, administrative and binding costs) amounted during the last years from 4.5 % to 3.5 % of the total university budget, being high by all standards.

The system provides the Library Centre an opportunity to comment on departmental budget proposals. The comments mainly bear on the consequences on personnel and space. The general policy is not to allow higher book-funds than can be processed and shelved. In the end the University Board decides. Up to now it has decided in two ways. In the beginning the tendency was to allow greater book-budgets to the departments, while expanding library staff and reading rooms. Because of last year's recession, however, further expansion of staff became difficult so that the budget proposals of the departments had to be down-graded to meet the work capacity of existing personnel. The next step is the daily task of book-selection. This is principally a departmental responsibility. Undersirable fragmentation, however, is prevented by the fact that each department has appointed one senior faculty member having coordinating responsibility. He acts in the same capacity as the German *Fachreferent* and the British "subject-specialist." His duties, as a contact person between teaching staff and library, approach the job description given in the British Report, cited above, par. 238.(2) Many librarians feel that this "faculty-orientation" endangers a coherent collection policy. Certainly there is a danger of inconsistency, of neglect of certain important fields, of overemphasis of personal concerns. In general, librarians have been very negative in their appraisals of the endeavours of non-librarians. (3)

In any case there are ways of counteracting these presumptions. One of them is the control by other members of the department over the activities of their selector. This control is quite effective because

1. all books are in open access and an unbalanced acquisition policy would show very rapidly in the book-stacks,
2. the Library Centre issues monthly lists of departmental acquisitions which are being sent to all department members.

On the other hand, the system of selection by members of the department has some advantages:

1. There exists no dichotomy between the aspirations of the department and library policy; in general the needs of the users are well provided for and anticipated; this is so because also the students are represented, and quite vocal, in each departmental assembly;

2. the library or the library system is considered to be an integral part of departmental activities;

3. the system effectively discourages the formation of institute libraries. The last fact must be stressed, for even if the selection by library staff were proven to have better general results, it would still be useless in case more than half of the book-funds were expended without any library control.
The question arises whether a long-term acquisition policy is necessary, desirable and workable for the total university system, in other words, for all sciences and for all forms of publication. In général it seems fair to say that this is indeed the case. The further question may be raised, however, whether an active, systematic purchasing program must be maintained for all subjects being taught and studied at a given university.

To this question my answer would be: Probably yes for periodicals (because one has to reckon with the recent findings of bibliometry); probably yes for monographs in the so-called soft sciences and the so-called non-sciences as well (because it seems difficult to systematically anticipate the use and nonuse of monographic literature in these disciplines). But can the same be said of monographs in the so-called hard sciences? Here we would have to say: Probably not (because periodicals usually meet all, the needs of scholars in the front line of research; books on the other hand are required as text-books or surveys only, and can, more often than not, be acquired in the current market place if needed).

At UIA, there is an attempt to improve the quality of acquisition by systematically reporting the selection of duplicates. It is evident that in a decentralized selection system the amount of unwanted duplicates can be very high. At UIA it amounts to about 10 % of all requests. Duplicate requests are not refused but sent back to the initiator with the notification that the book is already present in his own departmental library, in another one, or on order. The decision whether to purchase it as a duplicate or not, is left to the departmental coordinator. Most behave very sensibly and refrain from ordering duplicates when possible.

A second aid to a more integrated approach was the creation of a union catalogue of all departmental libraries, and this from the very beginning. This union catalogue is produced on COM-fiches; it gives a survey of everything the libraries possesses, by author, by title and by subject, across the boundaries of each department.

We have gone a step further in the case of periodicals. Two years ago, and again this year, a union catalogue was produced of all current periodicals present in the three Antwerp university colleges and in the City Library. Altogether there are about 8,000 subscriptions. The union catalogue was computer-produced and one of the by-products of this process were overlap-statistics. The general overlap in the four totally independent Antwerp libraries amounts to 40 %. One wonders whether 40 % is to be considered high or low. In another Belgian university, having one central library and 200 quasi-independent institute libraries all responsible to a single authority, the overlap was 60 %.

From the same computerized data, profiles were made of the periodical holdings of each library, enabling us to find out the exact strength of each library. On a totally voluntary basis the four libraries are now weeding out unwanted duplicates and exchanging backsets of those subscriptions which have been cancelled, the aim being that at least one of the four libraries has as complete a periodical set as possible.

This operation yielded but a modest result. Only 10 % of the duplicate subscriptions have been cancelled. But even this low percentage will offset the total production cost of the union catalogue within a period of three years.

References

1. University Grants Committee: **Report of the Committee on Libraries,** London, 1967, p. 101–104.
2. Ibid., p. 67.
3. J. P. Danton: **Book selection and collections,** New York, 1963, p. 75; A. O. Kouwenhoven: **Wetenschap en bibliotheek,** Amsterdam, 1971, p. 207–217.

Cooperation between University and Library in the Area of Collection Development in "New" University Libraries in the Federal Republic of Germany

M. Pauer

Library systems, in which librarians and scholars cooperate in collection development, did not develop in the Federal Republic of Germany until about 1964, at the newer universities. Up to that time there had hardly been any models for this new type of university library in German-speaking countries. University library systems organized according to this new type have been established since then – I state the names roughly in chronological order – at Bremen, Regensburg, Constance, Bielefeld, Dortmund, Augsburg, Trier-Kaiserslautern, Düsseldorf, Bayreuth, as well as in the comprehensive universities of North Rhine-Westphalia, and a couple of other cities. These libraries show quite distinctive features in respect to size, number of students, type of buildings, number of faculties, collections to be incorporated, and specific organizational structures. I cannot enter into these particulars in this report. Instead I will try to give you an idea of the characteristic ways in which university and library staffs have come to cooperate in the majority of these recently founded universities. Having been chief librarian of the University Library at Regensburg for 12 years now, I am particularly well-acquainted with the new Bavarian libraries; it is for that reason that I will tend to take my illustrations and examples from this area. The newer Bavarian universities, that is to say the universities at Regensburg, Augsburg, Bayreuth, and just recently Passau, have been organized along similar lines. They closely cooperate under the Bavarian "Universities-and-Academic-Act", working under the supervision of the Director-General of the Bavarian State-Library. The University Library Centre at Cologne performs the same kind of coordination among several similarly organized library systems at comprehensive universities within the area of North Rhine-Westphalia.

Requirements and expectations

These library systems combine both the functions and the collections of two quite specific types of libraries, which had developed in the older German universities from the end of the 19th century onwards, that is, on the one hand, the departmental library in the faculties and, on the other hand, the university library as universal library and lending library in the university. The new library systems are expected to offer the distinctive services and qualities of both types of libraries through an improved organizational structure and at less cost. While library staff expects increased benefits in the areas of planning and subject coverage, and anticipates improvements by means of pooled funds and centralized operations, scholars and their staff, working in the fields of teaching and research, can be relieved from the growing duties involved in running and administering a departmental library. The trustees of the library anticipate economies by rationalizing the organization and by avoiding superfluous duplicates. The users of the library will have access to the entire stock, arranged in a clear manner. They will no longer face the difficulty of having to use differently organized libraries. By offering the entire collection in a single, centralized catalogue – which can be distributed in many copies – better information is gained. Steps to deprive the professor, who had formerly been in charge of the departmental libraries, of any influence on the selection and arrangement of their daily-needed tools and books, have run into stiff opposition. Thus there is the understandable and reasonable claim on the new library systems, and on their organizers, to make the books available not too far from the offices of the professors and to put the stock in a systematic arrangement; furthermore, to enable the professors to have their say, and to play an active role, in the building of collections within the library system, that is to say, the distribution of funds and the selection of titles are not left to the sole competence of the library staff.

The organization of cooperative processes

Pooling the functions of both types of libraries requires that the ways in which librarians and professors coordinate their activities be defined and organized. To an appropriate extent the other users of the library, the students in particular, are to be given a part in the cooperative process as well. Matters of fundamental importance associated with library cooperation can be dealt with by academic committees such as those which have evolved at specific levels within the university structure. This includes deciding on the overall acquisition policy for the entire system as well as for individual subjects and departmental libraries. Decisions of this kind are, as a rule, taken by the library committees of the university and of the faculties or departments. The latter, however, have not always formed such committees. Daily problems are to be settled by persons who maintain contact with one another, and who are authorized and qualified to take quick decisions. On the level of the faculty and the department, the person concerned will usually be the dean or the library-advisor, as he is called. Library systems, in order to be prepared for this kind of cooperation, have to be organized in accordance with subject areas and in the case of large libraries, according to grouped subject areas. Furthermore, for the major fields of learning subject specialists and group-directors are needed to run the departmental libraries, that is, the subsections of the library system. These librarians usually have as their counterparts departmental library-advisors, who act as the representatives of an appropriate university committee, and who have been authorized to settle day-to-day matters. Insofar as individual departmental libraries or departments have formed committees, the subject specialist will, as a rule, belong to them. The same applies to the higher committees at the university level which are called library committees or library councils. These consist of professors, of junior lecturers, of a few students, and, as a rule, also of the chief librarian, a further member of the library staff, and an administrative officer from the university's central administration. In most cases a professor will be in the chair, but also, in a couple of cases, the chief librarian.

The planning of requirements and the distribution of budgetary means

A large field of coöperation between individual professors, the departments, the library and its subject specialists has been opened up by the indispensable (and extensive) planning for the collections and their future demands. This planning has been the prerequisite to the financing of collection building activities by the trustees. Since there existed hardly any studies, or models and patterns, on the requirements of the new university, opinions as to the size of the required basic collection has been divided. Negotiations and discussions took months. Comprehensive statistical material drawn from the older faculty libraries, the university libraries, the book production, and the field of academic publishing have been used at Regensburg, Bielefeld and Augsburg, in order to gain reliable data and make a valid prognosis.

Not only matters of scholarship and research had to be taken into account but also the book provision for a varying number of students, and finally, the organizational and physical structure of the individual universities concerned. Developing these medium-term financial plans, not only the librarians but also the professors have gained important indications in terms of the development of their collections and of their future tasks. This planning was of great help in the necessary process of marking off the acquisition-areas between individual subjects and their adjacent and overlapping fields. By means of these negotiations excessive claims resulting from acquisitions beyond a specific subject — a common practice in many departmental libraries of the old type — could be reduced to a financially acceptable level. In most cases, however, the evolution of the library system has, from time to time, called for the revision of existing planning and for its adjustment in accordance with available annual funds. The library committees of the university proved apt at this task as well as at the distribution of the annual budgetary means. The proposals brought forward by these committees have, as a rule, been accepted by the authorities which decide and control these matters.

Procedure for book selection

The individual engagement of librarians and academic specialists is a decisive factor in the book selection process. In my experience, the subject specialists of the university libraries cooperate with the scholars of the departments in a couple of different ways. There are, to my knowledge, departmental libraries in which book selection is almost completely left to the subject specialist's discretion, whereas in contrast, there are cases, where the head of the department is keen on discussing in detail with the subject specialist any title to be ordered.

To my mind the vital point is that there must be a speedy flow of complete information between the acquisition department in the central library and the subject specialist in the department. Otherwise the subject specialist is not in a position to fulfil his function of supplying expert information and guidance to the scholars of his special field. The acquisition department, as a rule, distributes copies of the national bibliographies to the special fields and their subject specialists. The same procedure applies to second-hand and antiquarian book catalogues and to publishers' prospectuses. These informative materials are to be handed directly on to the persons concerned, unless the subject specialist selects certain titles on his own on the basis of mutual agreements with the department.

A comprehensive submission of titles from the learned literature, sent on approval, has proved to be an important instrument for the information of the subject specialists and the professors. The university libraries at Regensburg, Augsburg and lately also at Bielefeld, have begun to order on approval the new publications of about 100 to 150 national and international academic publishers. A specific bookseller places standing orders with these publishers. As soon as a book has been published, the bookseller receives a copy and instantly passes it on to the acquisition department of the university library. The library assigns relevant titles to the subject specialists, who either display the books in their offices or pass them on to the persons concerned, for inspection and approval. In this manner both the subject specialists and the professors are in a position to decide the possible selection of the book on the basis of a personal inspection of the work itself. As the library system supplies all sections of the university, there will only be a limited number of cases in which the returning of books to the bookseller proves to be unavoidable. That is why these consignments sent on approval have shown to be sensible and justifiable from the economic point of view, both for the bookseller and for the library. The advantage of the procedure is the fact that, on the one hand, books can be speedily supplied and will be quickly at the library's disposal, and, on the other hand, that the order decision is based on personal inspection and that a considerable group of persons is involved in the decision. This procedure, moreover, provides an important motive for cooperation between the scholars and the specialists, because both parties feel able to take part in the decision at any stage. As these approval orders include a number of about 100 to 200 publishers, a large proportion of the academic book production is acquired in this way. This procedure, furthermore, offers the advantage that the acquisition department is able to roughly sort out the books, and that certain fields of learning can be assigned continually to one and the same subject specialist. Thus overlapping orders in peripheral subjects are checked at this early phase of the book selection process, so that double-purchases can be avoided. The subject specialist, moreover, is in a position to explain his assignment of the book to its classified shelf and to reach a mutual agreement on this particular matter if duplicates should be demanded for different shelves. He may, however, order additional copies if the specific title turns out to be needed for more than one shelf on account of its subject matter.

The role of the subject specialist in the book selection process

Reports form the Constance University Library stress the point that subject specialists make book selection decisions in up to about 80 to 90 percent of the cases, and that lecturers decide up to 10 to 20 percent. As far as I know, inquiries of such a kind have not been conducted at other university libraries. To judge from the quotas at the university libraries in Augsburg and Regensburg, the ratio of independent decisions by subject specialists tends to be considerably lower there.

However, the difficulty, as I see it, is whether decisions which are influenced by a great number of factors should be left to a single individual. Considering the complexity of the informative materials, such as publishers' prospectuses and brochures, periodicals, catalogues and national bibliographies or books sent on approval, it would be difficult to decide to whom the responsibility for the selection of specific titles is to be assigned. This aspect of the matter, however, is surely only of secondary importance. The point I want to stress is that the decisions must be shared by the users of the library so that they do not feel excluded from the book selection activities in the library. As I see it, in the case of highly specialized fields and their publications, the cooperation of as large as possible a number of academic specialists in the area of collection development is required. The feasible number of subject specialists at any library – their number will certainly not go beyond 15 to 20 – will probably be insufficient to cover every field of learning. The essential element in the subject specialist's work lies in his qualification to organize a cooperative effort, to provide the basis for decisions through preliminary information, to awaken an interest in as many parties as possible in the collection development program, and to direct this interest purposefully to the needs of a systematic build-up of the collections. The subject specialist will take supplementary steps, particularly when at certain times and under special conditions the interest of the department is not sufficient to ensure the acquisition of an adequate as well as a well-rounded collection. Thus the subject specialist in the Bavarian library system is authorized, and even instructed, to decide on complementary orders charged to the central library's account. This applies, in particular, to fields which are not represented by a professorship at a given university. He is, moreover, obliged to take an active part in the acquisition of the lending-stock and the student's textbook collection, from the budget of the central library. For this reason he may decide on his own to order books which have been turned down by the department. In practice there is little dissension between the library and the scholar in these matters. If at all, it occurs with second-rate books demanded for extremely specialized fields. There are cases, however, in which the subject specialist must insist on a written statement from the chair to the effect that a title or a duplicate to be ordered is in fact indispensable for scholarship.

At times the acquisition department of the central library, which is obliged to promote a supra-departmental coordination in book selection, will contact the library-advisor of a specific department, or the chief librarian has to be called in. If these negotiations, usually conducted between the department's representative and the chief librarian, should come to nothing they are passed on to the library committee. Dissensions of such kind, however, have hardly occured in recent years, because usually a satisfying compromise could be reached. This at least applies to the new universities. Disputes not settled at this stage must be decided upon by the president of the university. Students using the library, users from the region and, naturally, scholars have the chance to influence the orders of the central library by suggesting titles, which, as a rule, are accepted.

Acquisitions and the classified arrangement

The basic requirement of successful cooperation in the new library systems is the classified arrangement of the collections, both in the open-access areas as well as in the stacks. The university library at Regensburg expects to hold about 900,000 books in open-access zones, in about 12 reading rooms, by 1980.

The older and less intensively used books, which are kept in the stacks of the central library, may be looked over and be used there at any time too, by the subject specialists and the scholars. Thus an overall view of the collections is easily obtained. Contacts established on this basis, between subject specialists and professors, in order to change and extend the classification system and in order to assign books in peripheral and overlapping fields, help in avoiding superfluous duplicates. The easy access to the collections enables the library to face vigorously any claims that faculty libraries be extended in the direction of small general libraries – a tendency probably not unusual in the departmental libraries at the older universities. Excessive claims, if voiced at all in a few isolated cases, will be disapproved of promptly by colleagues from other fields who feel restricted in the use of their own funds. By means of this free interplay and balance of forces, and the chance to obtain useful information by looking over the collections themselves the bodies of the university which are concerned with library acquisitions thus can obtain more easily a clear picture of the overall situation at their university, necessary to reach an adequate decision.

The catalogue and acquisitions

Electronic data processing combined with new data storage devices has provided increased potentialities for cooperative acquisitions. While the printing out of catalogues by means of the Computer Forms Printer had, even previously, enabled the library to multiply its catalogues for use in the university library and departmental libraries, and this to a serious extent, the newly found technique of distributing catalogue data on magnetic tape in microfiche format has created the opportunity to place catalogues even at the disposal of individual scholars. In the near future we will be able to make available to nearly every professor older catalogues which provide quite sufficient information for acquisitions. This will to an even larger extent, I trust, improve the exchange of ideas between subject specialists and scholars on the development and growth of the collections, because at that time the catalogues can be placed conveniently on a desk and used there. The classified shelf list on microfiche would enable the scholar to gain at all times a clear picture of the acquired, or still missing, items reflecting the latest research in his field. We just have entered upon the initial stages of these new developments. The successful use of the printed catalogues of the older of the new universities, in the collection building by the newest ones, entitles us to conclude, even at this early stage, that catalogues based on microfiche will render most valuable services in the area of acquisitions, the more so since they can be provided in large numbers and virtually without additional cost.

The Relationship between University Library and Departmental Libraries at "Old" Universities in the Federal Republic of Germany

H. Sontag

In recent years, German universities have undergone substantial changes in regard to internal structure and size. This has naturally created structural changes within librarianship.

At the older universities, the library scene was characterized by the university library in "competition" with the departmental libraries. A large number of libraries, differing in size and user capacity and all having different funds at their disposal, were operating side by side, without any coordination.

It was the intention of the legislators of most state governments to bring about a change. The regulations, concerning librarianship, were rewritten within the frame of the university laws. Due to the rights of sovereignty, the state governments alone are responsible for their own university laws. The responsibility for the various legal obligations was divided between the federal authorities in Bonn and the state governments. The state governments were assigned the task of cooperating to solve all questions concerning cultural activities.

It has been found necessary to limit the scope of the regulations passed, since complicated circumstances have made it practically impossible to reform by legal action alone. Unfortunately, each state has also passed its own "library laws". This does not mean that all libraries are moving in a different direction as far as library systems are concerned, it does however follow, that Germany has no unified library law.

Nevertheless, opinions held by German librarians concerning well-functioning library systems, are quite similar. They only differ in the question of how to reach their aims. It's quite obvious, that the type of library organization applied at new universities, is very much influenced by new systems. It is also self-evident, that the methods applied at new universities are not transferable to universities having long traditions.

Research(1) has proven that the advantages of scientific study in small faculty libraries, due to their private character, are now outweighed, in

general, by the disadvantages that result from their isolated position (within the university complex). Historical facts may elucidate the economic problems facing the Federal Government, with regard to investments necessary for personnel and non-cash resources. In 1893 14 German universities had a total of 114 departmental libraries. In 1967, the State of *Nordrhein-Westfalen* alone had a total of 559 faculty libraries with book-stocks amounting to 5 million volumes. On the other hand, the affiliated central libraries only had a total stock of 3.4 million volumes at their disposal. At one and the same time, 33 social science libraries were in operation in Cologne, independently of each other, and without connection to the central library. In 1968, the State of *Baden-Württemberg* invested 7 million marks for the upkeep of 521 departmental libraries containing book-stocks of 4 million volumes. The aforementioned sum was almost twice the amount invested in the university libraries. Scarcely one quarter of the total book-stock was catalogued in accordance with standard cataloguing rules. 319, of these 521 libraries, only had sufficient storage space for less than 5 years. In order to solve this problem, it was necessary either to permanently increase the storage space available to these libraries, discard some of the literature, or transfer some of the books to other libraries with all the consequences of recataloguing. This should be sufficient to characterize the difficult situation at that time.
It is uneconomical, and wasteful, to split a university's book-stock in such a way that several different lending and administrative departments become necessary. This is of even greater importance if we consider that book-production, book prices, and personnel costs are increasing rapidly. There is also the question of the increase in the number of students, and the increase in interdiciplinary research, to be taken into consideration. Further, the problem of untrained personnel cannot be overlooked.
In view of this, the German Research Society *(Deutsche Forschungsgemeinschaft)* decided to search for ways to improve the situation. In 1970, their own Library Committee published the so-called "Recommendations for co-operation between university libraries and departmental libraries". In the case of newer universities, it had been possible, from the start, to avoid the "side by side existence" of the university library and the faculty libraries, thus creating unified, well-functioning library systems. As for the "old" universities, one could only try to improve the situation through an increase in cooperation, resulting in a gradual change in the internal structure.

The main recommendations, which are reflected in university libraries' recent endeavours, are as follows:

- Extension of the university library as a lending library, as an information centre, and as the main centre of library coordination. The university library should also function as the main storage depot for literature not permanently required.
- All qualified library staff, within the whole university complex, should be tabulated in the university library's employment tables, and all personnel engaged in library work (throughout the whole university complex) should be placed under the supervision of the professional staff of the university library.
- Smaller stocks of books should be concentrated in order to form larger, more efficient, administrative units. At the same time, an increase in the rate of acquisitions should be taken into consideration.
- Techniques should be unified, especially those concerning descriptive cataloguing, under the direction of the staff of the university library. The assignment of call-numbers should be uniform.
- The acquisitions departments should be coordinated, so as to avoid multiple purchasing. The number of books purchased would thus be reduced to meet the actual requirements.
- Joint information centres should be created at the universities, including union catalogues of books and of periodicals.

These objectives cannot be reached through the kind intention of the participants alone. Legal postulates, such as fundamental regulations and statutes, within the framework of university legislation, need to be created to ensure a coordinated library system. A university library board ought to be responsible, among other things, for the assessment and readjustment of the literature supply. This board should meet at regular intervals to discuss fundamental problems. Steps would have to be taken, and recommendations made, to ensure the coordination of librarianship and the development of cooperative library systems.

As already mentioned, legal conditions differ widely from one state to the other. Some of the states have included relatively detailed regulations within their university laws. For instance, *Rheinland-Pfalz* (§ 54 HochSchG) and *Baden-Württemberg* (§ 14 HochSchG).

Other states, such as *Bayern* (Art. 22 Abs 5 HochSchG), *Nordrhein-Westfalen* (§ 38 HochSchG) and *Schleswig-Holstein* (§ 62 HochSchG) simply state that "The library is a central institution and part of the university. The library comprises all of the university's books. . ." The regulations within the university laws which apply to Hamburg do not, on the other hand, provide sufficient basis to enable structural changes to be enacted. The regulations only signify that "The Committee (for librarianship) should concern itself with the coöperation of all libraries within the university with the national and university library" (§ 36 HochSchG). In this case the regulations are not at all precise.

I would like to recapitulate, on the subject of legal provisions, by pointing out that coordination and the acquisition of literature stand in the foreground. The problems concerning organization are only vaguely specified. Supervision of full-time library personnel is usually assigned to the head librarian of the university library. Some states, and most librarians, expect centralization to improve coordination within librarianship. This should result in greater effectiveness in respect to personnel employed and in the use of non-cash resources.

In my opinion, the development of well-organized, unified libraries at the "old" universities would have the following advantages:

- It would be easier to introduce unified library guidelines, and to execute them in compliance with the requirements.
- By employing qualified, or at least trained, personnel the most important book-stocks would be cared for well.
- The introduction of data processing techniques would guarentee greater effectiveness through improved utilization of central and extraneous services.
- Superflous reprocessing of book-stocks, in the case of change of location (from one library to another), would no longer be necessary.
- Existing book-stocks would be easier to locate.
- Users would expect to find similar provisions. Ideally the conditions would be the same in all libraries.
- Concentration of book-stocks, to form larger libraries, would make it possible to handle larger numbers of users without having to increase personnel. The opening-hours could, through the introduction of shifts, be extended.
- Fewer libraries would make the coordination of various acquisitions activities easier. In fact, this would not be possible without effective coordination.
- By avoiding multiple purchasing, one would be able to economize. The funds saved could then be utilized to purchase other required literature. This means that budget increases could be limited.

- Library personnel would be better employed in accordance with their qualifications. The daily requirements could be better met by changing employees' assignments where necessary.
- It would be easier, and fairer, to assess the quality of work done if all comporable procedures were similar (e.g. comparison of duties in all lending departments). The opportunities for promotion would improve. This generally would create greater equality among the staff resulting in less social tension.

There would of course be some disadvantages as well. It is evident that, in the course of budget debates, the easiest thing to do would be to cancel certain larger items instead of less visible smaller ones. However, should this be done, the revenue authorities would act against their own intention of creating a lucid financial structure and well-organized library systems.

The university professors would have fewer books (e.g. reference books) directly at their disposal. This would compel faculty members to make more use of special libraries (departmental subject libraries). The unconventional acquisition, and use, of literature would hardly be possible. The scope given to the individual employee in the libraries could be limited by an increase in qualified supervision and by the type of administration.

It should, however, be emphasized that the university professors' freedom of decision, with regard to the acquisition of special literature, would in no way be inhibited. Neither the various kinds of projects, nor the attempts to centralize and unify library systems, limit this freedom of decision. In some of the legal texts, this point is recorded explicitly. The limitation of acquisitions is mainly dependent upon the funds available.

What efforts have been made by librarians to improve the situation at the "old" universities? :

The main attention was, and is, given to the creation of central information centres (catalogues, bibliographies, etc.).

Agreements between university library and departmental libraries concerning the concentration of acquisitions are very important. Periodicals were, in this case, given most attention. Arrangements concerning the acquisition of periodicals, or the publication of union catalogues of periodicals as were made, for instance in *Freiburg, Göttingen, Hannover, Kiel, Mainz, Münster, Tübingen* and *Würzburg,* have usually signified the first step towards cooperation. Further improvements in cooperation usually depend on local circumstances.

Freiburg(2) and Berlin(3) serve as good examples to elucidate the development.

The university laws in *Baden-Württemberg* apply to the University of Freiburg. (§ 14 HochSchG) specifies the following points:

1. A unified library system, having a library committee to regulate basic problems.
2. The university authorities have the right of literature selection while acquisitions should be mutually agreed upon.
3. The university library is the central library. This library should have an union catalogue for all book-stocks etc. within the university.
4. The head librarian of the university library is, as chief supervisor, the head of all library institutions. He is bound to follow the directives laid down by the library committee. He is the chief supervisor of all library personnel.
5. Laws should be passed in which individual regulations are specified.

The following steps have been taken at the University of Freiburg:

- The catalogues of approximately 130 departmental libraries are gradually being incorporated into a union catalogue (union catalogue of monographs, union catalogue of periodicals and the Freiburg periodicals register). On 31.8.1975, 78 of the 130 departmental libraries had been incorporated.
  The catalogue cards were changed to the international standard size. Future cataloguing is to be carried out in accordance with unified rules (RAK).
  Regular training courses are being organized by the university library. In view of these measures, it has been necessary to totally recatalogue the old stock (on 31.8.1975, 34 libraries with a total of 267 000 catalogue cards).
- Centralized duplication of catalogue cards in the university library.
- Concentration of the smaller faculty libraries to form larger departmental libraries (where possible before recataloguing). In this respect the smallest libraries are being overlooked. In most cases, the reorganization of the libraries administrations has been carried out under the supervision of the university library. In the process, the chemistry, biology and physics departments transfered, in full, the administration of their libraries to the university library. Several other libraries have turned over their administration to the university library.

– The following points should be considered with regard to the acquisition of books:

1. New periodicals, reprints and works costing more than 200 DM should only be purchased after a referendum.
2. Consultations about works costing less than 200 DM are voluntary:
   a) Regular sessions concerning acquisitions are held between subject specialists from the university library and scientists from the faculties (especially when the library concerned is a departmental library administered by the university library), or
   b) through individual consultations (especially in the case of smaller and specialized areas of learning). In these cases, it is usual either to use the telephone, or to give written information concerning the intended acquisition, or
   c) through general consultations in respect to concentrations of bookstocks and principles of acquisition (e.g. law, economics and philosophy).

It is, however, impossible,to reach complete agreement with regard to acquisitions in a library system comprised of more than 100 libraries.

In his report,(2) the head librarian at the University of Freiburg concludes that, through concentration and standardization of administration under the supervision of qualified library personnel, better results may be attained. Furthermore, he maintains that by correct assignment of library personnel, within the departments, the quality of librarianship may be greatly improved. Taking this into consideration, it is clear that available funds may be more economically employed. One hopes to make further gradual progress after the administrative regulations have been enacted. The support of the chancellor's office and the board of governors, in unison with the various institutions, is necessary.

At Berlin, the university law is regulated through interpretable skeleton agreements. They indicate that the Permanent Commission for Librarianship should deliberate the basic questions concerning literature provision (§ 14 (5) UniG). The agreements also stipulate that library service is divided among the university library and departmental libraries. The libraries are obliged to cooperate in accordance with unified administrative principles. These principles, together with the library regulations, are issued by the Academic Senate (§ 24 UniG). The Permanent Commission is composed of two university professors, one researcher, one student, one general member of staff, and the head librarian of the university library (a total of six votes). At Freiburg, the board is, on the other hand, composed of the president of the university, the chancellor of the university, the head librarian of the university library, five professors, one researcher, one student and a librarian (a total of eleven votes).

The provisions, pertaining to the State of Berlin, in principle allow for an organization similar to the library systems employed at new universities. It is a prerequisite that the views of the university administration, the university librarian and the respective scientific departments coincide.

The university library at the Technical University of Berlin was able to begin, as early as 1966, with the development of its library system. The reorganization and incorporation of the Departmental Library of Architecture was the first step in this direction. At present, 17 departmental libraries have been reorganized into departmental libraries (administered by the university library). Several of these 17 libraries have been concentrated to form larger administrative bodies. Apart from this, all of the university's documentation centres have been affiliated to the university library, thus forming a main centre of documentation. During the second world war, the library was totally destroyed. In the course of reconstruction, an union catalogue was established. This catalogue gives reference to all works acquired by the Technical University of Berlin.

Due to this development, the university library has secured a leading position in the provision of literature. The resolutions, passed by the Permanent Commission for Librarianship, and the support given by the university administration have been most important. This has generally had a positive effect upon library administration and on coordination in the field of acquisition. The academic senate has not, as yet, passed library regulations.

At present, there are only a few larger departmental libraries which are not attached to the university library, (those of economics, mathematics and statistics). The university library also endeavours to assist the non-affiliated libraries with regard to cataloguing. This depends, however, on the number of staff available. In all cases, where sufficient space is available, the university library attempts to concentrate the smaller faculty libraries to create departmental libraries under its authority. Apart from the problem of accomodation, the question of staff is a great impediment. The faculties seldom employ full-time library personnel, which results in staff shortages in case of library transfer and concentration. It is seldom that the university library is able to provide the necessary staff.

In some departments, the professors are suspicious of centralization, a fact which creates further difficulties. Fears exist, among staff members in some institutions, that changes of locality and administration as a result of amalgamation, could reduce their influence. It is further assumed that less advantage could be taken of the professor's expert knowledge, and that the members of the faculty could lose their influence in respect to book selection. It is also believed that the university library's suggestion to select books during library committee meetings (held within the faculties) is not acceptable. It is maintained that the presence of the subject specialists (from the university library) is neither expedient nor his expert advice advantageous. The opinion of some faculty members, with regard to centralized library administration, is that it is neither more efficient nor requires less personnel. It is, on the other hand, believed to be more bureaucratic. Mr. Kehr reports that similar opinions exist in Freiburg.

In view of this, much persuasive power is necessary to convince the opponents of the advantages of the new system. Prototypes and facts are necessary to demonstrate that their fears are unfounded. Resistance should be broken down because no departmental library (administered by the university library) can function efficiently without the cooperation of the faculty. Faculty members, and other members of the scientific establishment, should regard the reorganized library as their "own".

The present difference between the Technical University of Berlin and the University of Freiburg is in the number of centrally administered libraries. The final stage, as I conceive it, should be one in which the university library takes sole responsibility for the supply of literature.

Library service at the Technical University of Berlin should be divided between the main library (central library) and approximately 30 departmental libraries. The latter should, dependent upon fields of study and space available, be able to accomodate book-stocks ranging from between 15 000 to 150 000 volumes. In cases where scientific units have reference departments at their disposal, these should also be supervised by the departmental libraries. It is clear that all departmental libraries should be centrally administered.

The number of reference books should be limited in order to avoid the re-establishment of uncoordinated libraries. It is difficult to standardize the number of reference books permitted since the requirements are dependent upon the subject area (e.g. human sciences, technology). Several universities have limited the number of reference books to 200 volumes. The Technical

University of Berlin makes individual agreements with each scientific unit in respect to the number of required reference books. In most cases, the scientific bodies themselves conduct the distribution. The administrative expenses, and management, of the reference departments should be kept at a minimum; for instance, when several copies of a reference book exist, no entry should be made in the union catalogue.
There is a most important difference between the library administrations at Freiburg and at Berlin. At the University of Freiburg all library personnel has been placed under the supervision of the head librarian. At Berlin, the head librarian is only responsible for the library personnel which appears in the university library's employment tables. This makes it difficult to coordinate library activities (within the university) and to carry out supervision by qualified library personnel (e.g. the head librarian). Various committees have been formed to discuss the numerous problems involved, with the intention of finding an impartial solution. It is apparent that differences of opinion play a major role here.

With regard to the selection of literature for the departmental libraries, the University Library at Berlin comes to terms with the members of the faculty. The university library has, on the other hand, no influence over acquisitions made by other libraries (libraries within the university complex having no direct connection to the university library). At Freiburg, the scientific bodies have the general right of literature selection. They are, however, compelled to agree upon their intended acquisitions, that is to say, with each other, and with the university library. (In case of conflict, the university library has to be consulted.)

At the University Library at Berlin the acquisitions of literature for the departmental libraries is not yet fully centralized. This is due to the fact that certain problems concerning accomodation, personnel and general administrative procedure, still have to be solved. Most attention should be paid to the centralized acquisition of periodicals, since it is in this area that the greatest advantages may be expected.
I would like to end by making a few personal remarks. The administrative concentration of all libraries offers various proven advantages. The main advantages are:

- Better utilization of non-cash resources,
- increased opportunities to guarantee minimal service in the lending departments, even in the case of extreme shortage of personnel (e.g. sickness, vacation etc.),

- better-qualified information and consultation because of trained librarians,
- unified descriptive cataloguing,
- greater mobility of book-stocks (change of location),
- more accommodation through elimination of multiple copies. Within the coordinated library system only one copy of a data medium is, in most cases, necessary. The remaining copies can be transfered to a regional storage library.

The library employees have a wider range of jobs or positions to choose from. It is relatively easy to change from one form of occupation the another.

An increase in size does, however, create difficulties with regard to personnel control, and improved organization is most definitely necessary. Following disadvantages can occur:

- The personal interest in the work usually decreases in proportion to the size of the organization,
- it is difficult to motivate employees to carry out special duties,
- rivalry can occur between departmental libraries,
- readiness to assist users can also decline since, in larger libraries, the amount of work to be done in the lending departments usually increases.

Within the libraries, one should try to avoid the librarian's inclination to carry out all tasks to the point of perfection, otherwise the advantages gained through reorganization can, through perfectionism, be lost again.

Despite all the problems to be overcome, I still regard the attempt to coordinate and reorganize all libraries at universities as correct. One should consider that:

- The huge increase in the number of students has changed completely the situation at the universities,
- mass processing is easier to handle through the introduction of data processing,
- the "old" universities will also be compelled to take advantage of electronic data processing. This cannot, however, be effectively utilized until all libraries within the university complex have been reorganized to form coordinated library systems.

References

1. Kehr, Wolfgang: Der Aufbau eines Bibliothekssystems – zur Neufassung des Hochschulgesetzes (§ 14) – in Freiburger Universitätsblätter, Heft 42 (Nov. 1973), S. 41

2. Kehr, Wolfgang: Bericht der Universitätsbibliothek Freiburg über die Rationalisierung des Bibliothekswesens in der Universität (Vervielfältigtes Manuskript) Bericht an den Landtag Baden-Württemberg vom 15.09.75
3. Kaegbein, Paul, Malz, Georg, Sontag, Helmut: Zum Bibliothekssystem der Technischen Universität Berlin. In: Vom Strukturwandel deutscher Hochschulbibliotheken. Zeitschrift fü Bibliothekswesen und Bibliographie, Sonderheft 14. Frankfurt 1973. S. 180–204

B. NATIONAL COOPERATION IN COLLECTION BUILDING

National Cooperation in Collection Building – The Problem in The Netherlands –

P. W. van Milaan

We have, in The Netherlands, thirteen universities including the technological universities and these for agriculture and for social sciences. Each of these has its own central library in conjunction with a larger or smaller number of departmental or institute libraries.

Some of these establishments for academic teaching and research are more than 300 years old; others belong to this century while the very youngest was founded this year in Maastricht.

Not all these institutions cover the same areas of learning, and this applies also to their libraries. Certain specializations such as agriculture, veterinary medicine, or various language fields may be quoted in this respect.

There is, moreover, considerable overlapping between collections and between individual publications in more than one place. In addition to specialization and duplication, however, there exist many, more or less serious hiatuses. This is apparent from the many requests for publications which cannot be met by any one of the libraries. The percentage of works which are nowhere to be found in The Netherlands is increasing from year to year.

This fact can be explained largely by the unfavourable ratio of funds available for purchase to the evergrowing flow of new publications. Other countries are faced with the same problem, of course, but the way it manifests itself and the way in which it is approached, may differ considerably from country to country. Amongst others, the following points seem important in the Dutch context – the fairly extensive autonomy of individual universities and technological universities with respect to the selection of academic literature and secondly, the lack of national organizations such as a central national library, a national lending library, a legal depository (I must add here that the latter is now being established) and depot libraries for rarely-used materials. Important too is the fact that every member of the Dutch academic community is expected to be quadri-lingual so that, without dinstinction, Dutch, English, German and French publications must be considered for purchase. In our favour is of course the easy accessibility of the various establishments to one another and this because of the short distances and good connections between them.

The overwhelming production of new literature and the limited purchasing power are not, however, the only causes of the many "white spots" or gaps in our national collection of books and journals, and in the annual addition of new publications. I have already mentioned the other side of the picture, the problem of overlapping. Very often this is unavoidable because, for efficient teaching and research, many publications have to be available in more than one place at the same time.

But, this does not apply to all the literature which is procured in multiple copies. That is, quite often duplication is a consequence of a lack in coordination. The real causes, however, lie much deeper. On the one hand there is a lack of exact data on the nature, range and importance of both the overlaps and the "white spots", on the other hand the investigation into the requirements of library users is still in its infancy. Up to now, unfortunately, a planned approach to the supply of world literature has remained merely wishful thinking.

As early as 1965 the *Rijkscommissie van Advies inzake het Bibliotheekwezen,* the State Advisory Committee on Library Affairs (till September 1975) began to struggle with this problem. It did not limit itself to the purchasing policy of, and the supply of literature to, the libraries in Dutch universities. Other Dutch libraries which maintain significant collections of academic literature were included in this investigation as well. Originally, a cooperative plan for collection building in the widest sense of the word was considered. Soon, however, the discussion moved in the direction of a "Coordination Plan for the Supplementary Acquisition of Foreign Specialized Literature".

That was the title of a report published in 1974, intended mainly as a memorandum for discussion. By "supplementary provision of foreign specialized literature" was chiefly meant literature of every sort which, as a rule, is not normally purchased. However, a clear definition of what this literature was to embrace was never given.

It was thought, however, that such a supply of literature from abroad would contain a great deal of specialized matter such as regional or local history which, wrongly, is not normally purchased. Some improvement in the situation could be brought about, it was felt, by a distribution of subjects and/or regions to, say, thirty libraries. The library which would be allotted a certain subject field would than be responsible for acquiring as much as possible of the relevant literature within this assignment.

Other libraries would keep this in mind so that the purchase of unnecessary duplicates could be avoided. In this way many of the "white spots" could be filled even though extra money would, of course, be necessary to do so. How much this would amount to is, as yet, an open question.

One of the objections to the plan states that it does not include a calculation of costs. Another concerns the vagueness as to the nature and extent of the gaps which would have to be filled by the "supplementary provisions". It has further been argued that no investigation has been made into existing user requirements. If one makes the attempt to fill up all gaps, there is indeed a good chance that a great deal of literature would be acquired which no one will ever need. According to these critics the first step should therefore be the carrying out of an investigation into user requirements, only to be followed by the planning necessary to ensure a more complete coverage of specialized literature from abroad.

Other objections to the Coordination Plan as drawn up by the State Advisory Committee pertain to the need to further work out basic ideas, the lack both of suggestions for improving interlibrary borrowing and of alternatives against which the proposed plan can be measured.

As our collegue, Mr. de Groot, explained during a recent meeting of the Flemish Association of Library, Archive and Documentation Personnel in Brussels, a great deal of this criticism may be called premature. The Dutch Coordination Plan pretends in no way to be a complete solution which can be put into action immediately. The plan is offered as a model in which a number of financial and organizational propositions – although important in and by themselves – have only been presented as a basis for more detailed consideration, Mr. de Groot speaks in this connection of a first phase to be followed by a second and then a third phase. He also draws attention to the significant fact that the Dutch Coordination Plan must not be considered in isolation but as a part of a much wider programme for a long-term policy.

He refers here to the report on "Academic Libraries in The Netherlands" published in 1969 by the State Advisory Committee on Library Affairs. This report gives high priority to the drawing up of a coordination plan for collection building. Other priorities mentioned are a legal depository, the establishment of one or more depots for little-used publications and the problem of coordinated collection building within one university.

Mr. de Groot spoke to you about this very problem during a previous meeting of this congress and I do not, therefore, need to go into it again. We should realize, however, that both problems are linked closely to one another.

I shall confine myself here to these general remarks on the Dutch Coordination Plan. In the first place, it is still under discussion and it is difficult therefore to prophecy what the results will be and secondly, I have not been asked to speak here on the Dutch Coordination Plan as launched by the State Advisory Committee, but on the problem of national collection building as such.

I should point out, in this connection, that those who drew up the Coordination Plan did *not* take the purchase of new books and journals as their starting point but based it on the collections already in existence. An enquiry was carried out in order to find out what various individual libraries actually had in their possession. This was done not to discover which "white spots" existed but which "centres of gravity" – that is more or less complete subject collections – were available. These "centres of gravity" would, by preference, have to be considered as recipients of further increases, consisting of supplementary specialized literature.

In this way the problems which arise in determining and defining lacunae would, as it were, be gotten around without difficulty. It has been said, and I believe rightly, that there is little point in striving after perfection in collection building without taking the requirements of the most frequent users into account.

It seems, therefore, most sensible to try and gauge these needs. The Library of the Amsterdam Free University (our host at this congress) has attemped to draw up a collection profile for its own university based on the long-term requirements of its users. In imitation of a.o. the Northwestern University Library at Evanston, Illinois, in the United States, this library initiated an enquiry among its faculty members, asking them to indicate to what extent certain publications ought to be collected within various subject fields. A scale of five categories was adopted, ranging from "less essential" via "basic information material", "material for normal teaching and research" and "material for current and future academic research" to "material necessary to maintain present collections of national and/or international academic importance".

These categories were defined as follows: From the "less essential material" no more should be collected than 0–10 %, from the "basic information material" only 10–20 %, from the "material for normal teaching and research" about 20–40 %, from the "material that is important for current and future academic research" about 40–80 %, and from the "material necessary to maintain present collections of national and/or international academic importance" as much as possible, that is from 80–100 %.

In this context a "collection profile" does not refer to an analysis of existing collections but to a prognosis for future collecting activities considered desirable for teaching and research on a long-term basis.

This is of importance particularly to a number of the younger universities such as the Amsterdam Free University, the Rotterdam Erasmus University and the Tilburg Catholic University, for developments which have taken place within these establishments since the sixties, have resulted in their collection building activities branching out strongly in new directions. Because of this, new collections are being developed which differ greatly from the ones built up previously. In the older institutions such as at Leiden, Groningen, Utrecht and the University of Amsterdam, this trend is seen to a lesser extent. There collection building is largely an extension of what is already present. Not so in the younger universities where collection building has to adjust primarily to new developments which are taking place there, and this often in fairly rapid sequence. I refer here, in particular, to the creation of new faculties, sub-faculties and/or inter-faculties.

In this way the collections built up in Rotterdam and Tilburg, till the beginning of the sixties, were entirely limited to the field of economics, but new collection building activities have been extended to the other social sciences, including law, economic history, geography, psychology, philosophy and in Rotterdam even medicine.

In drawing up a collection outline such as the Free University has in view, a first step has in any case been taken in the direction of a more detailed analysis of collection building requirements.

As I have said, the lack of an analysis of this kind is one of the objections put forward by the users community against the Coordination Plan designed by the State Advisory Committee. It is not surprising, therefore, that the Free University Library urges all Dutch universities to draw up similar collection outlines based on local requirements for teaching and research.

By grafting the Coordination Plan on to centres of gravity, as reflected in such outlines, instead of on to existing collections, the plan would be brought far more into line with the requirements for present and future academic teaching and research.

In both cases the solution to the "white spots" problem in national collection building is sought in a distribution of tasks, in other words, in assigning to an individual library, or to a group of libraries working together, one or more particular subject fields. The instructions would be to build up as complete and representative a collection as is possible for that subject.

The idea is related to the system of *Sondersammelgebiete* developed in the German Federal Republic. It is, moreover, not impossible that one or more libraries such as those of the Royal Dutch Academy of Sciences, and the Agricultural University in Wageningen, would develop into central subject libraries on the German model. In any case, the State Advisory Committee obviously preferred a decentralized organization because it believed that too few points of contact are in existence in Holland to achieve one central supply source for the additional publications to be bought.

Nevertheless, people in Holland have been heard to say that there is more to be gained from a centralized plan, in which the task of providing additional literature is given to one or two central lending libraries, on the model of the British Lending Library.

Without denying the advantages of such a solution, one must emphasize that the possibilities of this being realized are much smaller in Holland than in England. In any case, we have no adequate national library although the Royal Library in The Hague does fulfil certain national requirements. These are however largely limited to the humanities and the social sciences. We have not yet obtained official approval for a legal depository, attached to the Royal Library, to house Dutch publications, although an attempt is now being made to get this started on a voluntary basis. Although the Royal Library carries out important work in the field of interlibrary lending, by keeping central catalogues of books and journals, this has not led to the exercise of more centralized functions such as the furnishing of a central lending library of specialized and/or academic literature, which is much in demand.

Finally I would like to return to our starting point – the "white spots" or gaps in the national collection of academic publications. The Coordination Plan aims at filling these gaps by assigning collecting responsibilities to the most suitable scientific libraries. To the "white spots" themselves no further attention is given. I, however, consider a further analysis of these gaps to be important if only to prevent purchases being made which are of no use to anyone. What is the nature and extent of these "white spots"? How can they be located and how can they be divided into distinct categories? Are they found in all subjects or are there obvious differences?

Even a superficial survey taught us that we should not only speak of "white spots" but also of "grey spots". It is only when no purchases are made at all in a certain field that we are faced with true "white spots". This may, for instance, occur with publications written in peripheral languages. Usually, however, some material of this kind has been bought and then we have "grey spots". According to whether more or less is acquired on the subject in question determines to which extent the grey varies from dark to light.

The motives leading to non-purchase or part-purchase can vary. I have already mentioned the influence of language barriers. In other cases the nature of the academic programme, or of the material itself, may lead to a selective purchase. Often, too, it is for purely financial reasons that a certain choice has to be made.

We may not, therefore, in all fairness, lump together all white and grey spots in our national collections. Much non-acquired material has little or no significance for teaching or research and is, therefore, quite rightly, not bought. But, in many other cases, it is definitely a question of omission. The many requests for publications, made via the interlibrary lending system, which return unsatisfied to the applicants, are a clear indication of this.

Undoubtedly there is still a great deal to be investigated in all this. It would be inappropriate for me to refer here to foreign publications in this area. As far as Holland is concerned, I should like to mention in particular the thesis *Wetenschap & Bibliotheek* (Learning and the Library) by Dr. A. O. Kouwenhoven which appeared in 1971.

In this work he presents most meritoriously an analysis of collection building as an indispensible link in the process of the pursuit of knowledge. His theoretical observations can be of use in the further development of the Dutch Coordination Plan. The same applies to a case study he did in order to link up his earlier work to the academic field of social geography. If we really wish to achieve a purposeful national collection policy, similar case studies would have to be done in the various different subject fields and from them the theory would have to be tested and, if possible, perfected.

Local Acquisition and Discarding Policies in the Light of National Library Resources and Services

D. N. Wood

This paper is all about cooperation and the extent to which it is feasible and desirable. In particular it is about cooperation between the university libraries and the British Library.

Cooperation in the library context is not new although in the past it has more often than not been concerned with providing a back-stop to local resources. This was in the times when libraries regarded it as a duty to provide from their own stock virtually all that their readers required and when it was almost regarded as an admission of failure to have to borrow from another library.

The realisation that a 100 % library was not only unattainable but unjustifiable and the idea that cooperation should play a significant role in the development of acquisition and retention policies are relatively new phenomena. Certainly the philosophy of *national* and *university* library cooperation has developed only very slowly neither party expressing much enthusiasm for the idea. In the form of the British Museum the 'national library' in particular did not want to know and of the situation at the turn of the century Richnell in the Donald Urquhart festschrift says "The provision of research facilities for the growing number of academic communities. . . . did not appear to be any concern of the national library and it was the subscription libraries most notably the London library that provided a back-up service to scholars all over the country".(1)

During the 1920s the universities increasingly recognised that they could not continue to provide satisfactory services solely from their stock and that there was a need for some form of back-up service. Again, however, the 'national library' was unsympathetic and a separate body the NCL was created to fulfil this role.

Even after the second world war the isolationist view of the national library prevailed and was reflected in the BML's unwillingness to act as the national centre for bibliographic control. Once again a separate body the BNB was established.

In the last 20 years several significant developments have taken place, which have led to a further softening in university attitudes on cooperation and a rethinking of the role of the national library in this context. These developments may be summarised as follows: –

1. The creation and rapid growth of the National Lending Library for Science and Technology which demonstrated that a fast and efficient back-up service based on a control collection of literature was a practical proposition.

2. The rapid growth of the literature and the equally rapid increase in the cost of acquiring it and storing it. Both monographs and periodicals show a 4 % annual rate of growth over the last 15 years during which time the average serial subscription has increased from £ 4.50 to £ 20.40 and the average cost of adult non-fiction monographs has increased from below £ 1 to £ 3.64. The combined effect of these changes has meant that the policy of building new university libraries and extensions to accommodate ever growing quantities of literature has been costing increasingly more both in absolute terms and as a percentage of the total cost of running a university system. Between 1964 and 1974 over 23 million pounds was spent on major library buildings and annual expenditure on acquisitions, staff and services rose from 4.5 to 20 million. One additional reason for these increases of course was the third development.

3. The rapid expansion of higher education following the Robbins Report.(2) Not only did the existing universities increase in size during this period but new universities were built and others created from the colleges of advanced technology. The present number of universities in the UK is such that it is completely unrealistic for each to attempt to acquire and preserve the world's ceaselessly growing output of worthwhile literature.

4. The publication of the report of the Committee on Libraries(3) -a committee established by the University Grants Committee under the Chairmanship of Dr Thomas Parry. In the light of developments outlined in 2 and 3 above this committee saw fit to stress the importance of both local and national cooperation and in particular to define what it saw as the role and functions of a truly national library. The committee obviously saw considerable scope for cooperation between university libraries and an efficiently organised national library.

5. The report in 1969 of the National Libraries Committee(4) -a committee set up largely in response to the Parry Report. This contained recommendations which by 1974 had resulted in the formation of the British Library consisting of the former library departments of the British Museum, The National Reference Library for Science and Invention, The National Central Library, The

National Lending Library for Science and Technology, The British National Bibliography and The Office for Scientific and Technical Information.

With the bringing into being of the British Library the stage became set for cooperation and rationalisation on a scale which would have been unthinkable and impossible several years previously. Although there are several areas which give scope for rationalisation and cooperation between the British Library and the universities, the two which concern us today are concerned on the one hand with acquisitions and on the other with discarding policies. In each case the most obvious form of rationalisation is for a university library to refrain from acquiring or keeping certain types of material because of its availability in or from the national collection. One might add that because of the general unwillingness of research workers to travel far or often to use material in reference libraries it will be the material available from or through the Lending Division that will most usually affect local policies.

Let us look first at acquisition policies. To begin with we have to recognise that there are certain materials (some serials for instance) that are known to be in heavy demand and that the local library must therefore hold. Similarly it is essential that the library should maintain a basic collection of reference works in each discipline which it aims to cover. At the other end of the spectrum there are books, serials and other materials which any librarian will recognise as low use material, unlikely to be requested in his own library and, in view of its availability from or through a central collection, not worth acquiring. Between these two extremes however there is a multitude of serials and monographs for which some use can be predicted but which may not be high enough to justify purchase.

Help in predicting demand may come from the users. For instance a university lecturer may be able to provide valuable guidance as to what students will wish (or ought) to use. Alternatively the librarian may wait for demand to make itself felt and then decide to buy, or he may look for a clue at the demand in other libraries.

Having established or predicted demand he must then compare the cost of buying, processing and storing the item with the cost of obtaining it on inter library loan whenever it is required.

An interesting approach to the problem of rationalising library acquisitions was described by Houghton and Prosser in a recent paper in Aslib Proceedings. (5) They looked at serial holdings in industrial libraries and developed two methodologies for determining which serials were earning their keep and which were not. The first, based on ideas presented by Brooks in 1970(6) and which was very similar to a study carried out Williams et al(7) in the United States in 1968, looked at the *cost per use* of individual titles in comparison with the cost of obtaining a loan/photocopy from Boston Spa. The paper lists the operations involved in processing and handling literature and in obtaining material on inter library loan and costs there. It concludes that in the libraries examined the average cost for the former (excluding purchase cost) is £ 7.80 and for the latter 80p. The former figure is then used to determine the cost per use which can then be compared with the price of obtaining the item from BLLD. The inference is that where the cost per use is greater than this price the subscription could be discontinued. The authors themselves point to draw-backs in using this methodology without regard to other factors. Firstly, average processing costs are of doubtful value since the number of issues per year varies from one journal to another with the result that both binding and processing costs will also vary. Secondly, any library must cater for minority interests among the user population. Thirdly, since the price of journals varies greatly from one subject field to another the application of the above technique might lead to collections biased towards those areas served by relatively cheap publications.

In view of these limitations a second approach was suggested by Houghton and Prosser. In this they ranked the library's journals in descending order of use and produced a cumulative table by adding each journals usage in turn. The same was done for subscription rates and by plotting one against the other a graph of cumulative use against cumulative cost was constructed. It was found that the curve rose steadily and eventually reached a point where a fall off was observed. The law of diminishing returns was seen to operate at between 60 and 70 % of total journal costs and approximately 90 % of use. The implication is that those journals which represent 30 % of the acquisition cost but which are responsible for only 10 % of the demand can be disposed of. This appears on the surface to be a neat and practicable approach to the problem of identifying journals not worthy of their keep. However, as Line pointed out in a letter to the Editor of Aslib Proceedings(8) it should be realised that "the reduction of expenditure on journals can be the first step on a slippery slope. Since rankings by cost per use usually follow an exponential curve it can be shown that the librarian can reduce his expenditure by say 20 % while reducing the number of uses of the collection by only say 5 %. The deleted titles are then removed from the rankings which still of course follow an exponential curve. The following year a further 20 % can be cut from the reduced budget while loosing only 5 % of uses of the reduced collection. After a few years of this the librarian will be spending very little on journals. The total uses made of his collection will have dropped only gradually but the percentage of total requirements of his users that can be satisfied from his collection may well have fallen to an absurdly low level."

There are a number of other points which should be borne in mind when considering how the cost-effectiveness of one's library service might be improved by borrowing rather than buying.

Firstly, by failing to acquire low use material a libraries browsing and serendipity value is greatly reduced.

Secondly, it must be realised that availability affects demand. One would like to know how much additional use would be attracted by a serial or monograph which is asked for only occasionally on inter library loan were it available locally and of course vice versa.

Another consideration might be the extent to which particular titles were covered by bibliographies. For instance a borderline journal which was well covered by abstracting service might be dispensed with, while one used slightly less but not covered by such services might be retained on the grounds that if the subscription was terminated the information which the journal contained would become virtually inaccessible.

A further factor and one which bedevils any considerations regarding the cost-effectiveness of library and information services is the value of information. Houghton's studies were concerned with the cost per use. What we really want to know is not the number of times a title is used but the value of the information obtained from it. As Houghton himself says "the value to a researcher of one item picked up from a journal in the library may pay for the annual subscription to that journal in terms of the effect on the research workers productivity". If the journal had not been in the library at the time the article may never have been discovered.

I should now like to return for a moment to the cost calculations which may be used to determine whether to buy or borrow. Among the costs to be taken into consideration if the BLLD is used is the cost of the loan/photocopy form. Let us for the sake of argument say this costs 50p. The calculations may show that it is more economical for a particular library to borrow rather than buy a certain item. However, in national terms this may not be so. Consider for example an item wanted 5 times in its life and which costs £ 9 for a library to buy, process, store and issue. Assuming it costs £ 1.50 per item to borrow (£ 1 local costs and 50p BLLD form) it would be cheaper for the library to rely on inter library lending rather than purchase. However, the true costs of an inter-library loan are not £ 1.50. Local library costs of £ 1 may be accurate but the average cost of supplying an item from Boston Spa excluding acquisition, accommodation and overheads is around £ 1. There is thus a subsidy of around 50p per item borrowed. If the real cost of the operation were compared with the price of the book is £ 9 for the book and £ 10 to borrow it would be cheaper to buy. Thus by subsidising the price of forms the BLLD is encouraging, in national terms, uneconomic practices. If in the name of economy there is going to be an increasing reliance on the BLLD the question arises as to whether the loan/photocopy form should gradually be increased in price to a level where the subsidy is removed. There are other arguments for full marginal cost recovery although there are short term reasons against any sharp increase.

So far I have been assuming that rationalisation will take the form of local libraries curtailing *their* acquisitions and relying more on the national collection. This will usually but need not necessarily be so. There may be certain types of material where demand is so low and expertise in handling it so important that it is unnecessary or even impossible to develop and maintain more than one expensive collection. In such instances it may be appropriate for the national library to in effect delegate some of its responsibility for national provision to an academic or other type of organisation. Three libraries of the university of London were picked out in the Dainton Report for special mention in this context, they were the London School of Economics, the School of Oriental and African Studies, and the Institute of Advanced Legal Studies. How these and other libraries can be exploited effectively from a national view point and what financial support if any they could expect from the BL provides much food for thought. Certainly the BL is empowered to give grants to other libraries to enable them to develop para-national special collections or to exploit more full the collections they already have. In such cases however the problems which can arise due to conflicting local and national interests should be appreciated by both parties.

If university and national libraries are to be responsive to each others ideas and needs and if rationalised acquisition policies are to emerge it is important that they work closely together. One way in which this is being achieved in Britain is through the committees which advise the various Divisions of the British Library. These consist of experts among whom are representatives of different types of libraries and their function among other things is to in-form the BL of the problems and needs of local libraries and their users. Further feed-back on the effectiveness of the BL collections and services is obtained by BL staff who participate regularly at meetings such as this, and respond as far as possible to the expressed needs of individuals and user groups. It is important that specialist groups in their turn formulate their requirements with care and thought.

I now want to turn my attention to another related area where there is scope for university and national library cooperation. This concerns not the material which might or might not be acquired but that material which having been acquired might in the light of the low use made of it be discarded.

Apart from acting as a back-up for the material which becomes no longer available locally the national library can play an active role in the development of discarding policies by agreeing to accept and make available the material discarded by other libraries.

Although the size of library is frequently equated with the academic standing of the institution to which it is attached, the time has surely come when it must be realised that university libraries and indeed other types of local library cannot go on expanding at an ever increasing rate. As far as the UK is concerned the present is a particularly appropriate time at which to consider the whole question of discarding policies since many British university libraries are virtually full and this at a time when the Government is attempting to reduce public spending on all fronts.

One reason for libraries not having followed active discarding policies in the past is that de-acquisition is not a cheap process and it has been argued that it is better to spend ones limited resources adding material to stock rather than on weeding which is often difficult to do and can be politically damaging. With buildings not available however librarians may have no other choice in future. Not that this need necessarily dismay them. It need not be regarded as a criticism of past selection if a book acquired 10 years ago or even 2 is no longer required. Certain books particularly in rapidly developing fields go out of date quickly and once use has fallen off to such an extent that it cannot justify its place on the shelf it would be better in national terms to dispose of it to a central collection, for the statistical chances of such an item being wanted again in the library that bought it are much less than the chances of it being wanted elsewhere in the country. Conversely for low use material the chances of a wanted book being in the local collection are less than the chances of it being available in another library and if all libraries adopt a disposal policy the gain the advantage of being able to use the fast centralised borrowing system rather than having to first identify a location and then borrow directly from a library not geared to such inter library lending operations.

With the adoption of a disposal policy there would also be a national saving in space since the aggregated demand is likely to be so low that the national collection could safely keep only one copy.

There will be an additional benefit to the original library namely that from the time it disposes of the item the cost of lending it to other libraries will be borne by the national system.

How far should discarding go? Some have argued that it should take place at the same rate as acquisition thus giving rise to the finite library. This might be an attractive idea to the University Grants Committee but is impractical in the face of an increasing user population and an increasing rate of output of literature since, under these conditions, no library can remain the same size and still provide the same level of service. Some growth must therefore be allowed for. Nevertheless I can see no alternative but for discarding operations of some magnitude being mounted in most university libraries in the next few years.

The business of what to discard of course is not easy. Enright has suggested in his amusing paper on what he calls "Bibliochlothanasia" that of a university library's total stock some 15 % of it might be sufficiently heavily used to justify retention without question.(9) Similarly some 15 % of it might be so little used that its immediate discard would raise no problems. The difficulties arise in the remaining 70 %. Because of the uncertainty attached to any decisions regarding the value of individual items within this 70 % and because of the impossibility of arriving at any consensus among the library and academic staff any relegation would have to be on a trial and error basis. Perhaps what is called for is a local high-density low-cost store where material considered for relegation could be housed for a number of years while still being available within say 24 hours. If after say 5 years the material had not been used it could be permanently relegated. If demand justifies it individual titles or serial runs could be recalled to the main site at any time.

So far we have been looking at the problem of local acquisition policies in economic terms. What of another factor in the equation? How is the behaviour of a library's clientele affected by a reduction in the amount of literature available locally?

To answer this question we need to know, for example, how the extent of use of the library is affected by its ability to satisfy requests locally. It may be that below a certain satisfaction level use drops off alarmingly. It would be dangerous therefore to pass this point. Line has suggested that 95 % of undergraduate demand should be met by local library facilities or by a student himself and that 75 % to 80 % of research demand should be satisfied locally.(10)

It would also be desirable to know to what extent users are prepared to use alternative sources of supply. Are they sufficiently well informed about inter library services and their efficiency, to use these, or do they (as suggested by a survey at Birmingham University some years ago)(11) wander out of the library and take no further action if the item they want is not on the shelf?

Do users have sufficient knowledge of the guides to the literature for these to be an effective replacement for the shelves and catalogues as a means of locating references? Even where such guides are available, and many libraries have woefully inadequate collections of abstracting and indexing journals, it is doubtful if the user population has this knowledge. Any increasing dependence on inter library loan must therefore be accompanied by improvements in the holdings of secondary literature and by an active programme of user education.

Perhaps the most important question is what is the effect on the work of users of having to wait for items to be obtained on inter library loan?

On this point, Houghton and Prosser(12) in a follow-up paper to the one referred to earlier, looked at the effect of such delays on users in special libraries. Questionnaires were sent out to users with items which had been obtained for them from the BLLD. Respondents were asked to indicate the subject field in which they were working, the length of time that they had been waiting for the item, whether their work had in any way been held up while awaiting the arrival of the loan or photocopy, the significance of the information which was contained in the document and the aspect of their work in connection with which the paper or book was required.

Despite the wide range of waiting periods, users within the libraries surveyed did not appear to be inconvenienced by having to wait for material to be obtained from Boston Spa. Only 10 % of respondents indicated that their work had been held up. Furthermore, Houghton and Prosser found no correlation between the subject discipline of respondents and the degree of inconvenience caused by the delays. It would be interesting to carry out similar studies in other types of libraries particularly university libraries.

In some ways of course a finite library or one that grows only very slowly above a certain size may actually be *better* from a users point of view. A library which has more literature than it needs can be difficult to use, since the stock is widely scattered, the catalogue becomes unwieldy, browsing is less productive as readers are forced to scan shelves of relatively worthless and in some cases positively misleading works, valuable space is devoted to low use stock instead of seats for readers and staff are less helpful than they could be because of the dissatisfaction which they might feel from working in an over-crowded environment.

If the British Library and in particular the Lending Division was to be the final resting place for relegated material – and this seems not unreasonable – then resources for additional buildings and staff at Boston Spa would be necessary. However, this would be offset many times by the savings in the universities.

Whether we are talking about acquisition or retention policies one thing is clear and that is if the universities are to be persuaded to take national library services into account when developing their own policies then they must be persuaded that the national service is a *viable* alternative to local provision.

As far as the national lending service in Britain is concerned I think most university libraries regard this as adequate. The service based on the collection of the Lending Division of the British Library and backed up by the resources of other libraries is now dealing with around 2.5 million requests per year i.e. around 75 % of all inter library requests in the United Kingdom. The aim of the Lending Division is to acquire those categories of material which are either in substantial demand or which are difficult to obtain elsewhere. In terms of volume the most important holdings are serials and some 46,000 titles are currently received. These cover all subject fields and languages and include for instance over 3,000 cyrillic titles. A further 5,000 titles are on order. In the monograph field BLLD is collecting all "worthwhile" English language books excluding fiction but it collects foreign books on a very selective basis only. British Official publications are acquired comprehensively as are publicly available reports. Other types of material held include British and American dissertations, translations and conference proceedings. Details of holdings and current intake are shown in Table 1. Date-wise the collection is variable; heavily used serials for instance are held retrospectively in some cases back to the 17th century whereas the scientific monograph collection is only strong after 1960 and in the case of monographs in the social sciences and humanities comprehensive coverage goes back only a few years although the NCL brought with it a substantial if uneven collection of older books (some 600,000 pre-1960 books).

Table 1: Some statistics regarding BLLD stock

	Holdings	Current intake
Serials	100,000 titles	45,000 titles
Monographs	1,200,000	70,000
Reports	1,000,000	100,000
Translations	275,000	30,000
Conference proceedings	50,000	6,600
Dissertations	140,000	40,000

Besides providing a service from stock the BLLD provides access to literature held by other libraries. It does this via its Union Catalogue of Books (in which are recorded older English language monographs and foreign language monographs including many of the holdings of many universities), through microform versions of the catalogues of Manchester and Birmingham universities which are held at Boston Spa, and through various other services e.g. published catalogues of individual libraries, and by speculation with special libraries. Thus where an item is not available at the BLLD a location may be provided, thus allowing a requesting library to try an alternative source.

BLLD's ability to satisfy requests has been further improved by arrangements which have been made with a number of special backup libraries. These include the Royal Society of Medicine, the London Library, the various copyright libraries and last but not least the British Library Reference Division. The latter is prepared to provide photocopies of articles in periodicals held in the Department of Printed Books, the Department of Oriental Manuscripts and Printed Books and in the Official Publications Library. Unsatisfied requests from Boston Spa are also sent to the Newspaper Library at Collindale.

Finally and as a last resort unsatisfied requests for foreign language monographs may be sent abroad.

With this system of supplying items from a central loan stock, providing locations or passing material to backup libraries the BLLD is able to satisfy a large proportion of the requests it receives (see Table 2).

Table 2: A breakdown of requests received by BLLD and the ways in which they were dealt with. The data was derived from a study carried out in May 1974. The sample consisted of every 6th request received during 10 working days.

Category of material	Sample composition %	Total requests received	Requests referred back to requestor	Total valid requests	Valid requests satisfied									
					Supplied from BLLD stock		Locations supplied		Sent to back-up libraries		Sent abroad		Total satisfactorily dealt with	
					n	%	n	%	n	%	n	%	n	%
Monographs	22.7	3,065	415	2,650	1,615	61	688	26	133	5	39	1	2,475	93
English language	20.6	2,790	381	2,409	1,552	64	610	25	93	4	16	1	2,271	94
Foreign language	2.0	275	34	241	63	26	78	32	40	17	23	10	204	85
Serials	66.8	9,008	659	8,349	7,630	91	95	1	255	3	39	<0.5	8,019	96
English language	55.8	7,527	537	6,990	6,480	93	77	1	170	2	26	<0.5	6,753	97
Foreign language	11.0	1,481	122	1,359	1,150	85	18	1	85	6	13	1	1,266	93
Conference proceedings	5.8	785	133	652	530	81	33	5	5	1	3	<0.5	571	88
Reports, Official publications, etc.	4.0	535	75	460	385	84	19	4	3	1	4	1	411	89
Dissertations	0.7	95	21	74	54	73	1	1	0		10	14	65	88
ALL CATEGORIES	100	13,488	1,303	12,185	10,224	84	836	7	396	3	95	1	11,551	95

Being able to provide access to material is of course only part of the story. Access must also be rapid. The latest survey carried out at Boston Spa on this subject showed that around 50% of the requests which are being received are being dealt with in less than a day and that nearly 90% are being processed within a working week.

How quickly backup libraries are able to deal with requests will be looked at from time to time once this element of the service is well established. By paying backup libraries for providing the service however it is hoped to ensure that it is effective.

Could the national library services and in particular the services available from the BLLD to universities be improved? The answer is almost certainly yes. With the present acquisitions policies the satisfaction rate from stock should improve as time goes by and the speed of turn round of requests at BLLD is under constant study with a view to reducing the delays.

Significant improvements to the service could be achieved with the making available on loan material housed elsewhere in the British Library. There are obviously certain types of material which could not be lent from the Reference Division – British monographs, rare books, manuscripts, and so on – but there seems some scope for investigating whether there are some categories of literature which could be lent. For instance, as the present Director General of the Reference Division has himself said "There are extensive categories of foreign material acquired by purchase and gift of which the instrinsic value is relatively low and for which the demand in the reading room is so infinitesimally small that the loan of them would not only leave the reference potential of the library virtually undiminished but would provide a better justification for their acquisition and storage".(1)

If the Board of the British Library were to make policy decisions which would lead to the availability of reference material through the lending service then the Library would be in a much better position to rationalise its acquisition policy. This could lead to an even wider range of materials being available to the scholars and research workers of the United Kingdom.

These then are some of the ways in which the services available from or through the Lending Division might be improved. If improvement does occur it will lead to an ever increasing confidence in and reliance on the national library services.

At the same time it is only fair to try and identify developments which could lead to a deterioration in the service. In the first place the British Library cannot consider itself immune from the cuts which are taking place in public expenditure. Some reduction in the library's estimates seems highly likely although one could make the plea that if other libraries have to do with less there is even more reason for ensuring that the central collections are able to provide a full service.

Another question mark hangs over the postal service. At the moment this is the basis of the inter library lending operation in the UK and the BLLD postal bill alone at current prices is around £ 500,000. We are all aware of the deterioration which has taken place in the postal service in recent years and if this continues it could bring the inter library lending service into disrepute. However, alternative methods of moving photocopies and loans around the country are already being examined. A report sponsored by the British Library Research and Development Department has recently been completed by the Local Government Operational Research Unit. This looks at the economic viability of running a national road/rail transport system for inter library loans and the results suggest that it would be possible to organise a service which would be cheaper and no less efficient than the present postal service.

What is likely to have a more important effect on the efficiency and cost of running and using a central collection of literature is the outcome of various official investigations into the whole question of copyright. Among those things under investigation will be the effect of library photocopying operations, particularly operations on the scale practiced by the British Library, on journal sales. This topic could be the subject of a separate paper and much has indeed already been written. Publishing bodies have certainly been exerting a considerable amount of pressure in appropriate quarters with the aim of bringing about a reduction in the photocopying activities of libraries. If they were successful and photocopying was for example made illegal or prohibitively expensive what effect would this have on libraries such as the BLLD. One result would be to increase slightly the cost of inter library loan transactions through the increased costs of sending complete volumes and issues of periodicals through the post. Secondly, the service would become slower on two counts. Parcels take longer to deliver than letter post and unless a completely effective duplication policy could be devised long waiting lists would develop.

There are a number of reasons why at Boston Spa we feel that publishers are exaggerating the conflict between their interests and ours and our main arguments are outlined in a recent article in *Journal of Documentation.* (13)

In conclusion I would repeat what I said earlier, namely that there has never been a more opportune or necessary time for cooperation between the national and local libraries. I would urge university libraries not to regard the British Library as a competitor which could be used by fund-giving bodies as an excuse for giving less money to universities. It is a fact of current life that the sort of money we should have to develop and house our collections is just not available. Cooperation therefore is the only way forward and one can expect therefore that the philosophy of central collections providing backup services for local libraries thus allowing them to develop rational acquisition and discarding policies will loom large on the British Library scene for some considerable time.

References

1. Richnell, D. T.: The British Library and the universities: **In** Barr, Keith and Line, Maurice (eds) **Essays on information and libraries** London, Clive Bingley, 1975. 129–150.
2. Committee on Higher Education: **Report of the Committee on Higher Education** London, HMSO. 1963. (Command 2154).
3. University Grants Committee: **Report of the Committee on Libraries** London, HMSO. 1973.
4. National Libraries Committee: **Report of the National Libraries Committee** London, HMSO. 1969. (Command 4028).
5. Houghton, B. and Prosser, C.: Rationalization of serial holding in special libraries **Aslib Proceedings,** 26, (6), 226–235, 1974.
6. Brooks, B. C.: Photocopies v. periodicals: cost effectiveness in the special library **Journal of Documentation,** 26, 22–27, 1970.
7. Williams, G. et al: Library cost models: owing versus borrowing serial publications CFSTI, PB182304, 1968, 161p.
8. Line, M. B.: **Aslib Proceedings,** 26, (11), 447–448, 1974.
9. Enright, B. J.: Bibliochlothanasia: library hygiene and the librarian: **In** Barr, Keith and Line, Maurice (eds) **Essays on information and libraries** London, Clive Bingley, 1975. 61–78.
10. Line, Maurice B.: Local acquisition policies in a national context: **In** Jeffreys, A. (ed) **The art of the librarian** Newcastle Upon Tyne, Oriel, 1973.1–13.
11. Vickery, B. C., Wood, D. N. and King, M.: Report by Birmingham University Library on surveys carried out in 1964 on the use of the library by undergraduates, graduate students and staff.
    In University Grants Committee, **Report of the Committee on Libraries** London, HMSO, 1967, 213–228.
12. Houghton, B. and Prosser, C.: A survey of the opinions of British Library Lending Division users in special libraries on the effects of non-immediate access to journals. **Aslib Proceedings,** 26, (9), 354–366, 1974.
13. Line, Maurice, B., and Wood, D. N.: The Effect of a Large-scale Photocopying Service on Journal Sales **Journal of Documentation,** 31, (4), 234–45, 1975.

National Cooperation in Collection Building: Belgium

S. M. Namenwirth

I have been asked to explore with you the subject of national cooperation in collection building in Belgium. I must admit that I view this assignment with some trepidation and this for two reasons: In the first place, I have lived and worked in Belgium less than three years and therefore cannot possibly be considered an expert on matters Belgian. Secondly, whereas a great many things can be enjoyed and admired in Belgium – and it does not require three years to discover that – cooperation is certainly not one of them. In fact, very little has been achieved in the area of library cooperation thus far – particularly on the national level – and the prospects for future development are not bright. I could list the realizations here and leave it at that. It seems to me, however, that the Belgian situation is sufficiently interesting to justify a more careful analysis. If you allow me, I will therefore paint the backdrop against which Belgian cooperative efforts must be viewed; later I will discuss under which conditions a greater degree of coordination might be achieved. Which then are the factors inhibiting the cooperative process in Belgium?

No doubt you all know that Belgium is a bilingual country, with a touch of trilingualism. To librarians who tend to have an international – or should I say polyglot? – state of mind it must appear strange that multilingual conditions should affect the will to cooperate. In discussing these matters with my Dutch friends I always notice a certain incredulous disdain and I must say that I myself, when coming to Belgium, believed the language barrier to be a formal rather than a substantive issue, an added charm and challenge rather than a basic obstacle; I was clearly wrong.

Let me illustrate this point by describing to you the workings of the *Conference of Belgian University Librarians.* This body, sponsored by the *National Fund for Scientific Research,* represents a careful mixture of language groups, containing an equal number of Dutch- and French language members. When the president is Flemish, the vice-president must be Walloon. Every representation to the outside world must again be carfeully balanced. Whenever funds are available, they are divided strictly on a fifty-fifty basis. Although the discussions, generally speaking, take place in a friendly atmosphere, each member tends to speak his own language, and whereas the Flemish members can at least understand their Walloon counterparts, the opposite is not necessarily true. One might wonder why French is not used by all, since everybody present knows that language to a greater or lesser extent. However, given the long history of social, cultural and particularly economic domination of the Dutch-speaking population by the French-speaking upper classes, and given the long and often painful process of emancipation which followed and to some extent persists today, it is not surprising that Flemish librarians tend to insist on speaking their own tongue. I do the same but this practice does have real and significant consequences. The necessary presence of translators, the use of earphones, all help in creating a psychological barrier and tend to inhibit an open discussion of the give-and-take variety. It is interesting in this context that the *Convent of Dutch-Language University Librarians,* a group consisting of Dutch and Flemish librarians, meets in a basically more relaxed climate than the mixed-language Belgian Conference, although the informal, less official character of the international group may in part account for this fact.

In order to alleviate these difficulties, Flemish university librarians recently began to meet as a sub-group of the Belgian Conference. These meetings are indeed less formal but whether they will be more productive remains to be seen. This is so because many factors, other than linguistic ones, play a role in barricading the road to effective cooperation. Again for historical reasons Belgians view their country not so much as a single nation, or even as a federation of French- and Dutch-speaking sections, but as a conglomeration of individual towns and regions. The basic allegiance seems indeed to be regional rather than national; this being reinforced by the persistent and widespread use of regional dialects. It is not surprising therefore that cities such as Liege or Gand – each having long and proud traditions of regional identity or even independence – possess universities which tend to be particularly self-reliant, or if you wish, stand-offish. Discussions often reveal a greater concern for institutional autonomy than for solving common problems. Added to this, and probably related to it, is the individualistic character of the Belgian citizen. To illustrate this point, when dark sets in, the rolling-shutters are let down all over the country, thereby closing each and every family off from the outside world into a castle-like state of privacy. Finally, Belgian society is highly structured, and this not only linguistically and geographically but also culturally and politically. These very same divisions permeate university life as well, explaining the existence of state and free universities; the latter dividing into religious (Roman Catholic) and sectarian institutions. It is clear that this state of affairs can lead to a certain rigidity which in turn tends to foster the self-reliance I referred to before. Thus library careers tend to be intra- rather than interinstitutional; librarians tend to manifest institutional rather than professional commitments. It goes without saying that all these factors add up to a situation in which an effective system of national cooperation is hard to achieve.

Let us now look at the cooperative projects in the realm of collection development which have been instituted. On the national level there is in fact just one such program. The *Conference of Belgian University Librarians* – the august body I referred to before – created a program for the acquisition and dissemination of foreign periodicals. Participating libraries were assigned very broad subject areas – such as English-language science and technology – and were expected to select titles which were nowhere to be found in Belgium at the time the project was conceived. The underlying assumption was that the material would be sufficiently substantial to warrant purchase. In any case, no pretense existed as to the comprehensiveness of the coverage. The *National Fund for Scientific Research* made an annual subsidy of approximately $ 100,000 available to finance the purchase and binding costs of the nine participating libraries. It goes without saying that these moneys have been distributed according to a most complex formula although the full amount has never been exhausted.

At the urging of the *National Fund* the project is now being evaluated. Since the results are not in, it would be premature to make any formal pronouncements about its effectiveness. It would be safe to predict, though, that some of the periodicals are used lightly, that most of them are not used at all. This should not surprise us, considering that the program deliberately directs itself towards rarely-used materials; towards those periodicals which were considered insufficiently substantial to be purchased from standard resources. Yet, in the list of selected titles great qualitative differences can be detected, ranging from the decidedly second-rate but still perfectly respectable to the thoroughly trite and peripheral. It is not hard to recognize that some libraries gave priority to safeguarding certain minimum standards (and thereby failing to expend available income), whereas others did their utmost to use up the unexpected windfall (thus allowing themselves to fall below an acceptable quality norm). All in all, I cannot escape the impression that the program was instituted in order to spend available francs, rather than to meet a serious national need.

Yet, such a need does exist. Despite a number of richly endowed libraries and well-focused collection development programs, it is no easy task to perform systematic, book-oriented research in Belgium, particularly if the research falls within subject areas that are not specifically Belgian in scope. The reasons are clear enough: On the one hand the overwhelming production of scientific publications, on the other hand limited financial resources. On the one hand improved techniques of attracting attention to new sources of information, on the other hand insufficient library coordination, leading to both unnecessary duplication and unwanted lacunae, leading in short to an unproductive expenditure of available means. But whereas improved coordination in the collection development area seems definitely in order, most Belgian colleagues would agree with me, I believe, that the first priority in Belgium ought to be the improvement of the interlibrary lending and borrowing system, not the upgrading of collections. The *Center for Scientific and Technical Documentation* – a unit loosely associated with the Royal Library – maintains a union list on magnetic tape of foreign periodicals held in Belgium but no such tool exists today for monographs. Until something is done about that, interlibrary traffic will remain a slow and cumbersome affair. Also, until that time no cooperative scheme in the acquisitions field is likely to be very effective.

A cooperative project for the creation of a shared cataloging system and a national union catalog of monographs exists in fact, and although this matter falls somewhat outside of the scope of this paper, I take the liberty of saying a few words about it nevertheless. Financed by the *State Secretariate for Scientific Policy,* present plans call for the realization of an on-line development. The necessary analysis is being carried out by teams of specialists from the *Center for Scientific and Technical Documentation* and a number of Belgian universities. The make-up of these teams necessarily reflects the peculiar Belgian context, i.e. language parity and therefore language confusion. Furthermore, in order to create the most perfect possible institutional and lingustic balance, each team consists of two researchers and two administrative supervisors. It is clear that the great number of teams, and their particular composition, make an effective integration of elements very difficult. It is a highly ambitious undertaking indeed and one cannot be certain at all that anything real will ever come of it. At the same time, realizing its importance, Belgian librarians are spending a great deal of time and effort in the hope that the project will come to fruition despite these obstacles.

Now what about additional schemes in the area of collection development? A great deal of specific information is obviously required about the present state of affairs before anything further can or should be done. At this point too little is known about the scope and quality of existing library holdings and about the collection development policies that are actually in effect. A thorough study should therefore be undertaken to gather this information. From those data conclusions could be reached about present and future levels of duplication and omission. Once that stage has been reached, the same basic questions would have to be asked which confront you all, or have confronted you in the past.

First of all, assuming that a greater degree of coordination were necessary – and that adequate support could be obtained to do something about it – the question must be raised whether this ought to be done on a national, regional or international scale. From my earlier remarks it should be clear that every national project in Belgium has to cope with certain built-in difficulties. This fact must be faced realistically. An alternative would be the regional route. The librarians in the province of Antwerp, for example, have achieved a great deal, and this through initiative and good sense, not because of large sums of extra money. Thus a beautiful union catalog of serials available in the province is being published annually. The regional interlibrary lending and borrowing system has been improved – this in part because of the very existence of a reliable, up-to-date catalog – and although efforts towards a better coordination in the collection development area have been less successful, there seem to be realistic opportunities on the regional level. On the other hand, it goes without saying that regional programs, having limited ojectives, can but attain limited goals.

What about international cooperation? I must admit that I am rather weary of cooperative projects on an international scale. I am not referring here to the limited but important existing services: My own library for one could hardly survive without access to the resources of the National Lending Library in Boston Spa, or individual university libraries in West Germany or The Netherlands. I am talking about the development of formal cooperative schemes which necessarily would be tied to the cumbersome bureaucratic habits of international agencies. This does not exclude, of course, the possible creation of more limited, less official international cooperative arrangements. One such option, which Flemish librarians obviously have in mind, would be a closer working relationship with their Dutch colleagues. However, it is far from certain that anything viable could be achieved in this respect. Despite a common language, and geographical and cultural proximity, the force of habit and historically engrained national identities should not be underestimated. Remains therefore a national solution.

Despite what I have said before, something might be achieved in Belgium on a national level, as long as we do not fall into the trap of globalism. I very much doubt that the need exists, or the financial resources would become available, to create a Belgian parallel to say the NPAC or the PL 480 programs, or even the Farmington Plan. As I argued elsewhere, a small country such as Belgium should not seek to develop global programs; it should not strive for comprehensive coverage, neither in depth nor in scope. In other words, the cooperative plan ought to be reasonably selective, both in terms of titles chosen and in subjects or areas covered. Such a program ought to start from specific, identified needs rather than from some abstract, often grandiose concept. The purpose of a national program of this kind ought to be the meeting of a well-defined objective – namely the improvement of library service – not the fulfilment of a national dream; it should serve the library user, not considerations of national prestige.

Furthermore, no national program in collection development can be made to work unless an integrated, systematic policy has been developed in the general area of scientific documentation. At this point there exist in Belgium a host of agencies, committees and, if I am not mistaken, as many as seven official organizations in the field of librarianship and documentation. Until a single agency is created, having clearly defined responsibilities and adequate means to carry them out, no effective coordination of these multiple activities is likely. But, unless a national project in the area of collection development is part and parcell of an integrated program, covering such diverse matters as selection, acquisitions, cataloging, dissemination of information and inter-library loans, one may just as well forget about it.

If these conditions can be met, however, there still remains the question of a centralized versus a decentralized approach. The different models are familiar enough: A central universal library, a central lending library, central subject libraries or a diversified program involving all research libraries. Belgium is a small country and a single, centrally located library could, in principle, take care of most information requirements which transcend the abilities of individual university and other libraries. The Royal Library would of course be the most likely, in fact the only, candidate to take on such a function. On the other hand, the Royal Library, like so many institutions of its kind, is at least as much a monument as a library. It has significant custodial responsibilities towards the magnificent treasures it harbors; it functions as the legal depository of Belgian publications and issues the Belgian National Bibliography. It performs standard library functions too, of course, and especially the institutions located in Brussels, like my own, benefit a great deal from available resources and services. But the Royal Library is not a national library and lacks the means — possibly even the desire — to fill the existing leadership void; at this point it certainly fails in pulling together the many divergent tendencies in Belgian librarianship. I very much doubt therefore that the Royal Library can, or wishes to, function as a central universal library, either now or in the foreseeable future.

A central lending library (a library of libraries) may well be an excellent solution, generally speaking. It is highly questionable, however, whether the money would become available in Belgium to start such an institution from scratch. National subject libraries for such fields as medicine, agriculture or technology are also not very likely to emerge in the Belgian context, given the fact that at this time at least two major libraries do already exist for each of these subject areas; one in the Flemish, one in the Walloon part of the country, in addition to existing holdings in the Royal Library. Finally, what are the prospects for a decentralized scheme in the tradition of the Farmington Plan, the West German and the proposed Dutch programs? The experience with the foreign periodicals project, which we discussed earlier, seems to indicate that a decentralized approach can work in Belgium. It does in any case seem to reflect existing realities better than a centralized solution. A distribution of collecting responsibilities over different existing libraries, including the Royal Library, appears to be the most promising solution and does not necessarily eliminate the future development of national subject or lending libraries. As a matter of fact, I myself favor the most pragmatic possible approach; always starting from a recognized need, I would look for specific solutions to specific problems, thereby probably ending up with a mixture of existing models.

Dear colleagues, I hope you do not consider me presumptuous, having talked so long about so little. I do hope, though, that one of you — or should I say one of us? — will organize another conference on the theme of cooperation in collection building, say ten years from now. Maybe I will be invited again at that time and be able to report to you on a splendid cooperative project in operation in Belgium. However, I would like to go on record here that, unless such a program were actually in existence and flourishing, I would be obliged to decline the invitation. Thank you very much for your kind attention.

National Cooperation in Collection Building in the Federal Republic of Germany

D. Oertel

The efforts towards a better coordination of acquisitions and collection development activities among research libraries on the national level – or, as we say nowadays, on the supra-regional level – in Germany, go back to the end of the 19th century. At that time so-called "special subject areas" (*Sondersammelgebiete*) were assigned to certain Prussian university libraries by the Ministry, in order to establish a kind of cooperative acquisition scheme. During the twenties the *Notgemeinschaft der Deutschen Wissenschaft,* today's German Research Society (*Deutsche Forschungsgemeinschaft*), adopted the concept and continued the sponsorship of some historically grown centres of excellence. One example is the University Library at *Göttingen* which, due to its association with the Kingdom of Hanover and thus with England until 1837, has a precious collection of literature from and about Great Britain, and remains responsible for the special subject area England up to this day. But it is only since 1949 that a comprehensive cooperative plan evolved for the acquisition of foreign literature. The aim of the project was to make sure that, despite the very difficult economic conditions of those days, at least one copy of all important foreign periodicals and monographs would be acquired and made accessible through the interlibrary loan service. The instrument for its realization was the Cooperative Acquisition Programme of the German Research Society. In this Programme, the entire range of science – in the German sense of *Wissenschaft* – was divided into more than 100 subject fields of varying sizes, and responsibilities were assigned to 25 state and university libraries according to their traditions and capacities. In the beginning the Programme covered research literature in the classical formats such as journals and monographs only. The selection followed strict rules and was partly done by a committee of the German Research Society itself. The German Research Society bore the total cost of the acquired foreign literature, while the library, on the other hand, had to cover the expenses of German-language materials, and of binding and processing activities. Some measures for the filling in of gaps in library holdings since 1930 completed the Programme. Meanwhile things have changed. What had been created in 1949 in order to meet a serious calamity, has developed within 25 years into an instrument which allows the control of the present abundance of literature; an instrument which we now call "System for the supra-regional provision of literature". It comprises not only centres of excellence of different kinds but also tools for the location of, and procedures for access to, the acquired literature. Today the system is based on the following assumptions: On the one hand, the researcher must be confident that the literature he needs for his work will be made available to him within a reasonable period of time. On the other hand, not only the production of books, periodicals, reports etc. is expanding steadily, but there exists also a growing demand for specialized literature caused by the increased volume of research, the expansion of old and the foundation of new universities, the development of new disciplines and the improvement of documentation and information services. The demand, at the same time, directs itself to an evergrowing quantity of publications, from more and more countries. This situation makes it almost impossible to maintain, or establish, local autarchy in the provision of literature. And even a collaboration between neighbouring libraries is as a rule not sufficient to reach the needed capacity. In our opinion a national scheme is therefore indispensable to build a reservoir of publications, relevant for science and technology, research, administration and finance, and available in case of demand.

In order to create such a reservoir, two basic models are in existence: A centralized lending collection as represented for example by the British Library Lending Division, or a cooperative system as is functioning in Germany and is being considered now in Belgium and The Netherlands.

The memorandum of the German Research Society's Library Committee dealing with the supra-regional provision of literature, which some of you already know and which is in your hands in abstract-form, discusses the alternatives in one of its chapters.

Some points in favour of a centralized system are:

1. There is no problem of delimitation between disciplines; conflicts between local and supra-regional tasks within libraries are eliminated.
2. There exists one partner only and no complex regulations for interlibrary loan procedures are necessary.
3. The administrative and technical organization of such a centre can be geared towards the economic handling of interlibrary loan requests.

Advantages of the decentralized solution are in our opinion:

1. The library responsible for one or few disciplines can concentrate itself more fully on specific user demands.
2. A professional organization for the supra-regional provision of literature corresponds with a similar professional structure for scientific information services.
3. A collection of periodicals and series can be built up centrally with some success; this is not true however for monographs and peripheral literature (*Graue Literatur*), especially in the humanities.

Our decision to continue the decentralized system is based not only on practical, but also on political and financial reasons: The Federal Republic of Germany is constitutionally a federal state, and the responsibility for academic librarianship lies mainly in the hands of the *Bundesländer.* The establishment of a central lending collection in Germany is therefore not very likely, even if we should like to have one. We should further take into account that the Technical Information Library, the Central Library of Economics and the Central Library of Medicine, together already hold over 40,000 current serials and that the State Library of the Prussian Cultural Patrimony (*Staatsbibliothek Preussischer Kulturbesitz*) is subscribed to about 30,000 serials, all of which are available on interlibrary lean. Nobody in our country would favor the duplication of these holdings, particularly since these collections comprise in part rarely-used materials. On the other hand, it would be impossible to amalgamate existing holdings, since legally they belong to different owners and are under the authority of different administrative bodies. We hope to remove existing deficiencies in the system by taking measures compatible with the decentralized approach. The recent "state of the art", as well as expected developments in the German system, can be described as follows:

1. The centres of excellence collect partly "classical" disciplines such as archaeology, theology or medicine (subject coverage), partly countries, regions or complete continents as for example Great Britain, Scandinavia or Africa (area coverage). Together they cover in principle the whole spectrum of the "sciences". Area coverage pertains to publications dealing with, not published in, a given country or continent. Although it would be easier, from the librarians' point of view, to collect in one place all literature produced in say South America – from linguistics to technology –, selection according to discipline remains also in this instance the dominating principle in the German system. Problems do arise however with interdisciplinary subjects such as

ecology. On the whole, scientific developments have to be followed closely in order to perceive new tendencies and to avoid gaps within the system. The delimitation between the centres of excellence is ensured by rules formulated by the German Research Society.

2. Besides periodicals and monographs (documenting the results of new research), the importance of "non-conventional" scientific literature (e.g. reports) and other publications of a documentary and informative nature (offering facts, dates and opinions) is growing, as they represent indispensable research tools. The necessary expansion of the acquisition programme in this direction tends to overburden the "classical" special subject areas at universal libraries, the more so when it is combined with a growing demand for information activities concerning those publications. In the case of disciplines, or complexes of disciplines, where this problem plays a special role, so-called "central subject libraries" have been established which represent a high concentration of resources in their respective disciplines and which maintain comprehensive acquisition programmes. Such libraries now exist for technology, medicine, economics and agriculture. A similar solution is being considered for education.

In addition, many special libraries – mainly in the humanities – have been integrated into the system during the last years, and this with varying objectives. Moreover, there are efforts being made to expand or improve the traditional special subject areas where necessary. Some examples are psychology, linguistics and also theology. For example, some of the cooperating libraries are now publishing accession lists, bibliographies or current awareness services.

Last but not least I should mention the State Libraries in Berlin, Frankfort and Munich. While the *Deutsche Bibliothek* in Frankfort acts mainly as an archive for German publications and as a national bibliographical centre, the State Libraries in Berlin and Munich are fully integrated into the interlibrary loan service. On the national level, they form large "general background libraries".

3. In the acquisition policy of centres of excellence, the principle of selectivity has been modified in favour of a nearly automatic comprehensiveness, for example, by acquiring all periodicals which are analysed in the leading information service within a given field. The reason for this policy is our experience that titles are ordered through the interlibrary loan system without any regard to their individual value, as seen by the librarian. Such a global arrangement

is certainly not cheap but we believe that it can be justified in the case of central, national resources.

4. Library collaboration within the German system takes place on a voluntary basis. Planning and steering activities are carried out by the German Research Society, being a central organization for the promotion of research, especially basic research, in the Federal Republic of Germany.

In respecting the autonomy of these agencies which govern libraries, – first of all the *Länder* – the German Research Society is not concerned with local problems of individual libraries but with the supra-regional cooperation among libraries and among central institutions, and with models for new developments. One has to bear in mind that in the Federal Republic of Germany there does not exist a central, national library administration and that the federal government gives financial support to a few libraries only, such as to the *Deutsche Bibliothek* in Frankfort.

The German Research Society tries to fill this void to a certain degree; it is not itself part of the public administration, although it disposes of public funds.

Those libraries which participate in the system receive annual subsidies from the German Research Society, in compensation for the additional tasks which they undertake in the interest of the general public; mostly for acquisitions, partly for personnel and travelling costs.

As far as the special subject areas are concerned, there is a fixed distribution ratio between subsidies and library contributions, while the financing modalities for the central subject libraries are very different. In 1976 the German Research Society granted 8 million DM for the centres of excellence, not including additional grants for cataloguing activities which are also part of the system, as for example the union catalogue of periodicals maintained by the State Library in Berlin. In our opinion the combination of central planning and steering, together with supplementary central financing, is an essential prerequisite for the proper functioning of a decentralized system.

5. Three problem areas in our system, which are discussed over and over again, should not be concealed:

   a) Participating libraries may be faced with conflicts between local and supra-regional functions, especially if money becomes scarce.

   b) Although instruments for the correct transmission of loan requests do exist, mistakes occur repeatedly. In some cases it cannot be recognized at once to which centre of excellence a certain title belongs. This can cause annoying delays.

c) The Regulations for Interlibrary Loans (LVO) in the Federal Republic of Germany, issued by the *Bundesländer,* are based on the regional principle. This means that interlibrary loan requests normally have to pass the regional union catalogues and only in certain cases may go directly to libraries being responsible within the supra-regional system. In practice, however, many interlibrary loan requests are sent directly, so that the central subject libraries can maintain a direct service, functioning outside the normal interlibrary loan provisions.

In cooperation with the appropriate committee of the *Deutscher Bibliotheksverband,* the German Research Society is studying available options with a view to improving the present situation, hoping to be successful within the next years.

Developments in Collection Building in University Libraries: Resolutions and Recommendations

The following resolutions and recommendations were adopted at the final session:

1. The meeting believes that universities should ensure that their libraries' acquisitions should be made, and their collections deployed, in the most effective way for their users. Developments of recent years include:

    The growth of universities, and of student numbers;
    the development of inter-disciplinary studies;
    the continuing growth of the volume of academic publications;
    the increased need to make the most efficient use of library staff.

For all these reasons it is necessary to re-examine existing selection and acquisition policies.

It is *recommended* that all acquisitions should be made in accordance with an agreed general collection-building policy for the whole university.

2. The competent authorities in each country *should ensure* that each university library is provided with the resources needed to facilitate cooperation with other libraries and other institutions.

3. The competent authorities *should also ensure* that each university library is provided with additional staff and other resources to undertake promising cooperative projects, both within the university and beyond.

4. Individual libraries *should* undertake small-scale cooperative schemes on a pragmatic basis, to provide models for other similar schemes.

5. Those bodies responsible for planning new academic developments *should* consider from the outset the provision of adequate library resources for those developments.

6. In each country a study *should* be made of how a policy of national co-ordination in collection-building can be achieved. This study should be based upon an evaluation of the present situation.

7. Any policy of national or international coordination *must* include provision for a comprehensive collection of each nation's publications (including non-book materials), a law of legal deposit, and the publication of a national bibliography.

8. To achieve the goal of national and international coordination (i.e., the economic provision of the widest possible coverage of the world's literature, and its most effective use), *a pre-requisite* is the existence and maintenance of a speedy and efficient inter-library loan service, with minimum legal restrictions on the making of photocopies.

9. There is a need for a comprehensive system of regional and/or national union catalogues, or alternative forms of record, not only for inter-library loans but also for other purposes connected with co-ordination.

The Conception of the Free University Library

J. Stellingwerff

1. Universities in The Netherlands

The oldest university in the Northern and Southern Netherlands is the University of Louvain, now 550 years old. After the rebellion against the Spanish-Habsburg monarchy, Leiden became in 1575 the first and most influentual university of the Northern provinces. However, the other United Provinces liked to have their own university or *atheneum illustre* as well. When The Netherlands became a department of the French Empire, Napoleon, with one decree, made all the universities part of the Imperial University of Paris. Only Leiden and Groningen maintained their university status. After the defeat of Napoleon there were three universities, each having a few hundred students, at Leiden, Groningen and Utrecht. Just one hundred years ago the economic depression made it necessary for the Government to abolish the University of Groningen, but those who wished to maintain that institution and those who wished to reconstruct the old *atheneum illustre* of the city of Amsterdam into a new Amsterdam University, constituted the majority. So 99 years ago, instead of a reduction to two universities, we witnessed the extension to four. Three years after that, in 1880, the Free University at Amsterdam became the fifth university in The Netherlands and the second in Amsterdam.

The bill of 1 October 1877 made it possible for any Dutchman, any recognized society and any church to establish a university. Before that date a university was an organization of the state, now the first university at Amsterdam became an organization of the municipality and the second an organization of a free society. I am proud that my two grandfathers, a shoemaker and a draper, and my father, who was a teacher, belonged to that society, as I do.

2. The Free Reformed University at Amsterdam

Thus Louvain is 550 years old and the Free University at Amsterdam only 96. A century ago we saw in The Netherlands the struggle for emancipation of the Roman Catholics, of the working-classes, of the Jews, the first movement of women's liberation and the struggle for the emancipation of the low church people. In the nineteenth century the established reformed state-church increasingly became a liberal organization of the state. As a reaction against that development there were secessions in 1834 and again in 1886.

The first secession caused the 1846 emigration to the United States of America, to Holland, Michigan and Pella, Iowa. The second secession was intimately connected with the founding of the new Free University. The society which established the Free University consisted of low church people: presbyterian, puritan or independent Calvinists led by Mr. Guillaume Groen van Prinsterer and afterwards by Dr. Abraham Kuyper. So the Free University was a small, reformed or Calvinistic university.

Yes, a very small one indeed, for the University started with five professorships and five students. All the money came from contributions from the members, and from those who sympathized with the Free University Society.

In 1905 one of the founders of the Free University, Dr. Abraham Kuyper, was Prime Minister and through his legislation the School of Engineering at Delft became a technical university and the degrees of the Free University received formal state recognition. After the first world war all primary education – governmental and denominational – became subsidized by the Government. After the second world war the Free University and the Roman-Catholic University at Nijmegen received at first partial subventions from the State, complete ones since 1970.

In other words, during the first sixty-five years of its existence, the members of the Free University Society had to carry all the financial burdens of their university. And these same people had to pay for their new churches, for their own political party (the oldest in The Netherlands) and for other activities geared towards their emancipation.

Money was necessary for the salaries of the professors and for the maintenance of a double-house in Amsterdam, on the Keizersgracht. During the first fifty years studies were limited to theology, law, Hebrew, Greek and Latin, and history. There was no money for a library. The library of the University of Amsterdam functioned as a city library as well, and was furthermore located in the neighbourhood of the Free University. Therefore the Free University developed only a very small library during the first fifty years, built up from gifts and from the private libraries of deceased professors.

3. The Free University Library

In 1930 the study of the sciences (mathematics, chemistry and physics) became possible after the building of a new laboratory. And after the second world war a curriculum for economics, the social sciences, medicine and dentistry, biology and geology was at last added. During that period the Government began to grant increasing sums of money. When in 1960 I became head librarian of the Free University, the university had spent less on its library in 75 years than I had had available for the library of the new technical university of Eindhoven in three years.

In Eindhoven there had been about 30 staff members, in the Free University only five. The first thing I had to do was to make plans for a new library building, inheriting some fundamental starting-points: No separate institute libraries and a single, total library system without competition between professors and librarians. The total library budget had to be under control of the librarian and all the different buildings in Amsterdam had to be concentrated on one campus consisting of only four or five big buildings. There was no money for a double system, so library services had to be connected within each big building to the departments of learning. I had to group the institute libraries together in a number of different locations, and work them into a central library system. And last but not least, I had to build up a library such as the Free University did not have before.

In The Netherlands we do not have a system of colleges; the university works through faculties. The buildings of the Free University are concentrated in four units within a small area. The main building for the alpha faculties (theology, law, the arts and economics); a laboratory building for the natural sciences and mathematics; a laboratory building for the medical sciences together with the university hospital, and a gamma building, not yet under construction, for the social sciences. Four buildings, and thus four libraries, an alpha, beta, gamma and a medical library. Four libraries but one organization.

The main building of the Free University has sixteen floors, the library being on the ground floor, the first floor and the third to fourteenth floors. On the ground floor we find the central catalogue and the bibliographic room, on the first floor the administrative department and the central reading room. From the third to the fourteenth floors we have six duplex-units. A duplex is a combination of two floors. A duplex consists of one entrance with a loan desk, of two reading rooms, two rooms for subject specialists (or *Fachreferenten;* we say *wetenschappelijke bibliotheekmedewerkers*), two rooms for library assistants, and behind the loan desk an entrance to three levels of stacks and at the end of the reading rooms an emergency staircase.

The entrance to a duplex in on one side of the six main elevators, on the other side there is the building for the faculties. So there is a direct connection between the faculties and the library. The elevator lies between the two and forms the link between the faculties and the library duplexes.

In the stacks we have a permanent elevator for books; behind the loan desks there is an elevator for library staff. These six duplexes will eventually become the alpha-library. Temporarily one duplex and a half are used by the gamma library until a new building for the social faculties is constructed.

The beta library will be on the top floor of the beta building while the departments of mathematics, physics, chemistry, geology, biology will have their own departmental libraries underneath.

The medical library consists of about twenty institute libraries with one centre in the middle of the medical building and another in the hospital.

4. Policy for collection building

When we had to build these new library facilities, there was no money to fill gaps in the collections. We bought the library of an old Roman-Catholic seminary, we received the very rich and interesting library of the late Mr. Hendrikus Bos, and the library of Dr. Abraham Kuyper. So we have an interesting collection but not an all-round university library. I believe that we shall never be able to round out our collections, so that we shall have to adopt a different collection building policy. That policy was formulated in a *collection profile.* The latter contains a description of the goals of the collection development programme. It represents a cross-section of the collection, projecting in general terms the size and composition of future collections as needed for instruction and research at our university.

It is not an easy affair to develop a collection profile. We had to start by asking the faculties questions by subject groups. The results were coordinated and adapted by the subject specialists in the library. And now I am in the process of coordinating and adapting the results of the subject specialists in order to present that product to the library committee and to our library staff.

After that, the collection profile will be a very important tool in preparing our budget proposals. It will be the corner-stone of our library policy.

Our theoretical starting-point in making the collection profile is what we call the bibliographical universe, or the total world-output of documentary information (books, pamphlets, journal articles, reports, materials on microform and so on). From that you select what is needed for the library of your university.

The world-output of relevant, scientific and learned publications can be measured in numbers of bibliographical units. We use Unesco numbers for that purpose.

This world-output of publications has to be divided into languages and subjects. If you know the total production in a certain language and on a certain subject you can fix the percentage you have to buy for your own collections. In this manner you can build a collection profile and a necessary budget, if you know the average price per bibliographical unit.

It is only in the case of limited subjects and accurate bibliographical data that you can apply this method. Usually a more intuitive method has to be employed. In that case we do not use percentages but indications of degrees. We have chosen the following five categories:

Degree ZERO: *No collection building* but sometimes individual gifts or gift collections are accepted.

Degree ONE: *Basic collection,* that is a collection of manuals, reference works, important bibliographies and the most significant works as a stand-by collection or a collection for a subsidiary subject.

Degree TWO: *Normal collection* as needed for learning and research at our university. Such a collection is to be developed in all subjects of learning in case there is not a better and open collection in the surroundings.

Degree THREE: *Special emphasis collection,* a collection for special research and learning at the university.

Degree FOUR: Universal collection, that is a collection of virtually all literature on a subject.

To make a link between the two methods, we might suggest the following approximate distribution:
Degree zero, 0–10 percent;

degree one, 10—20 percent;
degree two, 20—40 percent;
degree three, 40—80 percent;
degree four, 80—100 percent. In other words,
Degree one means basic literature only;
Degree two a careful selection;
Degree three the flexibility to create a broader and really good collection.

The only similar collection profile I know is that of the Northwestern University Library at Evanston, a few miles north of Chicago, Illinois, U.S.A. That collection profile has been most helpful for the development of our own ideas. The five degrees chosen for Evanston are more comprehensive than ours. They are:

a. A basic information collection, appropriate in the subject which falls outside the scope of present instruction, yet within which readers may need minimum library service to aid their understanding of material properly within the scope of their studies;

b. A working collection, which is a collection adequate to convey existing knowledge of the subject in broad outline, including its main historical aspects, and thus sufficient for the needs of undergraduate courses;

c. A general research collection, which is adequate for the needs of graduate students in the subjects, since it includes the major portion of the materials required for dissertations and other independent research;

d. A comprehensive collection, which embraces all the materials of the general research collection, plus a wider collection of books and periodicals having value for current research, and additional works for historical research in the subject, in all pertinent languages, though not necessarily in all editions, or in translation, with inclusion of considerable documentary and original source material.

e. An exhaustive collection, which endeavors, so fas as is reasonably possible, to include everything written on the subject, in all languages, of all time, in all editions and translations.

5. National cooperation in collection building

A collection profile is not only a tool for the coordination of collection building within a university but also a tool for national coordination of collection development activities.

I believe that we have to build up a national library network consisting of scientific and learned libraries connected to a computerized central catalogue of all their new publications, together with a central catalogue in micro-form of older collections present in each participating library.

At that time all literature present in The Netherlands will be available to everyone on interlibrary loan. Each library will therefore have to make its collection profile. Afterwards all these collection profiles will have to be amalgamated in order to create a supplementary national collection profile.

Through this system of collection profiles we will know whether an individual title must be purchased or not; through the central catalogue system we will know if one or more copies of a book are already available in the national library system.

This is the way we are working in our library and I hope that my colleagues will start working in the same way.

That's my story about our university, our library and our ideas concerning the problems of coordination in collection building.

APPENDICES

Appendix I. Programme.

Opening of the symposium	W.R.H. Koops (Groningen)
Welcome speech	C. de Niet, Board of Managers Free University
Introduction	W.R.H. Koops
1st session	
speakers	K. Garside (London)
	J.R. de Groot (Leiden)
chairman:	H.D.L. Vervliet (Antwerp)
rapporteur:	G. Kissel (Bremen)
2nd session	
speakers:	H.D.L. Vervliet (Antwerp)
chairman:	B.M. Bland (Southampton)
rapporteur:	B. Adams (Bochum)
3rd session	
speakers:	M. Pauer (Regensburg)
	H. Sontag (Berlin)
chairman:	G.A. Hamel (Enschede)
rapporteur:	J.L.M. van Dijk (Maastricht)
4th session	
speakers:	P.W. van Milaan (Rotterdam)
	D.N. Wood (Boston Spa)
chairman:	R. Landwehrmeyer (Tübingen)
rapporteur:	W. Dieneman (Aberystwyth)
5th session	
Introduction to the library of the Free University	J. Stellingwerff (Amsterdam)
Visit to the library	

6th session

speakers:	S.M. Namenwirth (Brussels)
	D. Oertel (Bonn)
chairman:	N. Higham (Bristol)
rapporteur:	P.J.C.A. Pinxter (Eindhoven)

Visit to the library of the
University of Amsterdam

Final session

chairmen:	W.R.H. Koops (Groningen)
	J.R. de Groot (Leiden)
rapporteur:	E.R.S. Fifoot (Edinburgh)

Appendix II. List of Participants

B. Adams,
Universitätsbibliothek der Ruhr-Universität, Bochum, Federal Republic of Germany.

B. M. Bland,
Southampton University Library, Southampton, Great Britain.

A. C. Bubb,
Salford University Library, Salford, Great Britain.

T. H. Bowyer,
Queen Mary College Library, London, Great Britain.

D. A. Clarke,
British Library of Political and Economic Science, London School of Economics, London, Great Britain.

D. Cox,
Leeds University Library, Leeds, Great Britain.

W. W. Dieneman,
University College of Wales Library, Aberystwyth, Great Britain.

K. W. Drozd,
Staatsbibliothek Preussischer Kulturbesitz, Berlin, Federal Republic of Germany.

J. L. M. van Dijk,
Bibliotheek Rijksuniversiteit Limburg, Maastricht, Netherlands.

E. R. S. Fifoot,
Edinburgh University Library, Edinburgh, Great Britain.

R. Fulford,
British Library Reference Division, London, Great Britain.

K. Garside,
London University Library, London, Great Britain.

J. R. de Groot,
Bibliotheek Rijksuniversiteit Leiden, Leiden, Netherlands.

D. Grosheide,
Bibliotheek Rijksuniversiteit Utrecht, Utrecht, Netherlands.

G. A. Hamel,
Bibliotheek Technische Hogeschool Twenthe, Enschede, Netherlands.

N. Higham,
Bristol University Library, Bristol, Great Britain.

W. M. J. Jansen,
Bibliotheek Katholieke Hogeschool, Tilburg, Netherlands.

D. Jenkins,
Wales National Library, Aberystwyth, Great Britain.

G. Kissel,
Universitätsbibliothek Bremen, Bremen, Federal Republic of Germany.

W. R. H. Koops,
Bibliotheek Rijksuniversiteit Groningen, Groningen, Netherlands.

F. J. Kühnen,
Zentralbibliothek der Medizin, Köln, Federal Republic of Germany.

R. Landwehrmeyer,
Universitätsbibliothek Tübingen, Tübingen, Federal Republic of Germany.

A. J. Loveday,
Secretary Standing Conference of National and University Libraries, London, Great Britain.

P. W. van Milaan,
Bibliotheek Erasmusuniversiteit, Rotterdam, Netherlands.

U. Montag,
Bayerische Staatsbibliothek, München, Federal Republic of Germany.

S. M. Namenwirth,
Bibliotheek Vrije Universiteit, Brussel, Belgium.

D. Oertel,
Bibliotheksreferat, Deutsche Forschungsgemeinschaft, Bonn-Bad Godesberg, Federal Republic of Germany.

M. Pauer,
Bibliothek der Universität Regensburg, Regensburg, Federal Republic of Germany.

P. J. C. A. Pinxter,
Bibliotheek Technische Hogeschool Eindhoven, Eindhoven, Netherlands.

G. G. A. M. Pijnenborg,
Bibliotheek Katholieke Universiteit, Nijmegen, Netherlands.

C. Reedijk,
Koninklijke Bibliotheek, Den Haag, Netherlands.

H. Sontag,
Bibliothek der Technischen Universität Berlin, Berlin, Federal Republic of Germany.

J. Stellingwerff,
Bibliotheek Vrije Universiteit, Amsterdam, Netherlands.

J. Stoltzenburg,
Bibliothek der Universität Konstanz, Konstanz, Federal Republic of Germany.

J. S. Teague,
The City University of London Library, London, Great Britain.

P. E. Tucker,
Warwick University Library, Coventry, Great Britain.

H. D. L. Vervliet,
Bibliotheek Universitaire Instelling Antwerpen, Antwerpen, Belgium.

H. Vogt,
Niedersächsische Staats- und Universitätsbibliothek, Göttingen, Federal Republic of Germany.

G. Wiegand,
Universitätsbibliothek Kiel, Kiel, Federal Republic of Germany.

D. N. Wood,
British Library Lending Division, Boston Spa, Great Britain.

S. van der Woude,
Bibliotheek Gemeentelijke Universiteit, Amsterdam, Netherlands.

G. G. van Wijk,
Bibliotheek Landbouwhogeschool, Wageningen, Netherlands.

J. Zandvliet,
Bibliotheek Technische Hogeschool Delft, Delft, Netherlands.

INTERNATIONAL FEDERATION OF LIBRARY ASSOCIATIONS
Serial IFLA Publications
Edited by Willem R. H. Koops and Peter Havard-Williams

1 **Special Libraries – Worldwide.** A collection of papers prepared for the Section of Special Libraries. Edited by Günter Reichardt. 1974. 360 pages. DM 68.00. IFLA members DM 51.00. ISBN 3-7940-4421-5

2 **National Library Buildings.** Proceedings of a colloquium held in Rome, 3–6 September 1973. Edited by Anthony Thompson. 1975. 144 pages. DM 28.00, IFLA members DM 21.00. ISBN 3-7940-4422-3

3 **Le contrôle bibliographique universel dans les pays en développement.** Table ronde sur le contrôle bibliographique universel dans les pays en développement, Grenoble, 22–25 août 1973. Edité par Marie-Louise Bossuat, Geneviève Feuillebois, Monique Pelletier. 1975. 165 pages. DM 38.00, IFLA members DM 29.00. ISBN 3-7940-4423-1

4 **National and International Library Planning.** Key papers presented at the 40th Session of the IFLA General Council, Washington, DC, 1974. Edited by Robert Vosper and Leone I. Newkirk. 1976. 162 pages. DM 36.00, IFLA members DM 27.00. ISBN 3-7940-4424-X

5 **Reading in a Changing World.** Papers presented at the 38th Session of the IFLA General Council, Budapest, 1972. Edited by Foster E. Mohrhardt. 1976. 134 pages. DM 28.00, IFLA members DM 21.00. ISBN 3-7940-4425-8

6 **The Organization of the Library Profession.** A Symposium based on contributions to the 37th Session of the IFLA Council, Liverpool 1971. Edited by A.H.Chaplin. 1976. 2nd edition. 132 pages. DM 28.00, IFLA members DM 21.00. ISBN 3-7940-4300-X

7 **World Directory of Administrative Libraries.** A guide of libraries serving national, state, provincial, and Länder-bodies, prepared for the Sub-Section of Administrative Libraries. Edited by Otto Simmler. 1976. 474 pages. DM 60.00, IFLA members DM 45.00. ISBN 3-7940-4427-4

8 **World Directory of Map Collections.** Compiled by the Geography and Map Libraries Sub-Section. Edited by Walter W. Ristow. 1976. 326 pages. DM 48.00, IFLA members DM 36.00. ISBN 3-7940-4428-2

9 **Standards for Public Libraries.** 1977. 2nd corrected edition. 53 pages. Paper DM 12.80. IFLA members DM 9.80. ISBN 3-7940-4429-0

10 **IFLA's First Fifty Years.** Achievement and challenge in international librarianship. Edited by Willem R. H. Koops and Joachim Wieder. 1977. ISBN 3-7940-4430-4. In preparation.

11 **The International Federation of Library Associations and Institutions: A Selected List of References.** 2nd edition, revised and expanded. Compiled by Edward P. Cambio. 1977. VI, 51 pages. Paper DM 16.80. IFLA members DM 12.80. ISBN 3-7940-4431-2.

12 **Library Service to Children**
Edited by Colin Ray. 1977. app. 160 pages. ISBN 3-7940-4432-0. In preparation.

Verlag Dokumentation München – Postfach 71 10 09 – 8000 München 71

Bei Fragen zur Produktsicherheit wenden Sie sich bitte an:
If you have any questions regarding product safety,
please contact:

Walter de Gruyter GmbH
Genthiner Straße 13
10785 Berlin
productsafety@degruyterbrill.com